p. 1-25

STUDY GUIDE

TO ACCOMPANY

GEOGRAPHY

REGIONS AND CONCEPTS

FIFTH EDITION

HARM J. DE BLIJ

PETER O. MULLER

University of Miami

Prepared by

PETER O. MULLER

Outline Maps Designed by

GYULA PAUER

University of Kentucky

JOHN WILEY & SONS

New York Chichester Brisbane Toronto Singapore

ISBN 0 471 62599 X
Printed in the United States of America

10 9 8 7 6 5 4 3 2

A FOREWORD TO STUDENTS

This Study Guide is designed to enhance your learning of world regional geography, building upon your parallel experiences in the classroom and in reading the chapters of the textbook. It is recommended that you have an atlas handy; the base maps in the textbook will readily suffice if no other atlas is available. The only other resources you will need to complete the exercises in this manual are a set of colored pencils or pens and (optional) tracing paper.

One of the main purposes of this Study Guide is to get you to become more familiar with maps (a subject introduced on pp. 4-5 and 51-53 in the textbook). For each world realm we have provided four outline maps (located at the end of each chapter) on which you are asked to locate and enter important geographical information. The essay questions in each chapter are also attuned to maps--many, in fact, cry out for a cartographic answer. That is exactly what you should consider doing: use sketch maps whenever the map is the best way to explain what you mean (like a picture, a map is often worth a thousand words). You can use tracings of the printed maps in order to get the necessary outlines of the geographic realms; it is also a good idea to use such tracings in answering map questions as a sort of "test run" before placing your final answers on the printed map. Moreover, placing information on maps as you read the textbook can be a valuable technique for later reviewing before examinations; an extra printed outline map is always provided in each chapter of this manual. This map of Southern Africa shows what a typical review map might look like after basic geographic information from your reading and Study Guide exercises is added:

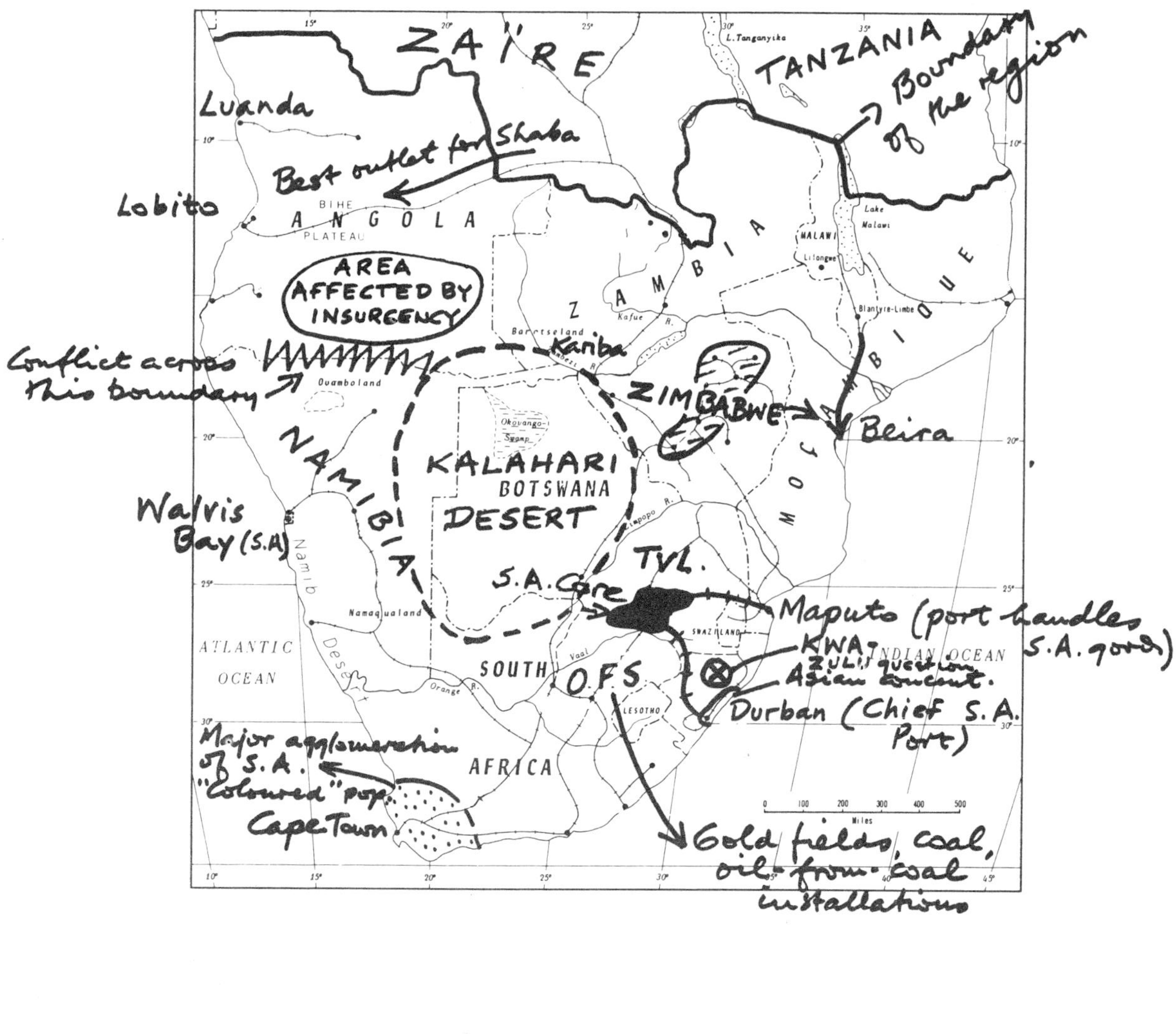

Each chapter of this Study Guide follows an identical format. The opening section describes the objectives of the chapter together with a list of things you should be able to do once you have learned the content of the parallel textbook chapter. This is followed by a glossary of terms covered in the book chapter, including appropriate terminology from the main textbook glossary as well as additional definitions; the Study Guide glossary is arranged in the order in which terms appear in the book and carries page references to the text. Then comes the main study and review section, a series of detailed side-by-side questions and answers for you to sharpen your understanding of the high points of the subject matter. Map exercises appear next, divided into two sections: (1) a series of map-comparison queries arranged around the textbook maps, and (2) three assignments to fill in basic physical, political-cultural, and economic-urban geographic information for each realm; as indicated above, a fourth blank map is included for your own use. After this comes a practice examination consisting of short answers (multiple choice, true-false, fill ins, and matching) and some longer essay questions--an answer key for the short-answer questions is found at the back of this manual. A term paper pointers section follows which includes tips on the preparation of research papers on the realm and the subfield covered in the chapter's Systematic Essay, and is built around a guide to the literature cited in each chapter-end bibliography. The final item is a special exercise that offers an enrichment experience, usually developed around a major idea covered in the book chapter (such as in the Model Boxes).

You should find these exercises helpful in this course, particularly for personal studying purposes. They will also assist you to learn basic world geography and to sharpen your map interpretation skills. Such knowledge is not only important as a contribution to your becoming a liberally-educated person--it may already be vital for success in the increasingly internationalized economy and society we all belong to.

A FOREWORD TO INSTRUCTORS

The aims of this Student Supplement to the fifth edition of Geography: Regions and Concepts are spelled out in the student foreword above.

This manual is designed to complement the text in several ways with each chapter providing a lengthy glossary, a detailed list of study questions-and-answers, map exercises that require comparisons and placement of spatial information on outline maps, practice examination questions, term paper pointers, and special enrichment exercises. You should also be aware of the availability of the Instructor's Manual to the fifth edition, which contains a number of items pertinent to the teaching of a world regional geography course based on this textbook.

ACKNOWLEDGMENTS

I am most grateful to a number of people for their assistance and advice during the preparation of this Study Guide. The manuscript was produced in camera-ready form by Fleur Sequeira, Roz Padilla, Debbie Mastrud, and Laverne Meister of Word-O-Matics in Miami. The maps were designed by Gyula Pauer, Cartography Laboratory Director, Department of Geography, University of Kentucky. At my department office, Lourdes Rivera, Ivonne O'Donnell, and Nancy Gutierrez performed countless supporting tasks with their usual efficiency and cheerful spirits. I also acknowledge again those who helped in the preparation of the previous edition of this Study Guide--on which this latest version is based.

Peter O. Muller
Coral Gables, Florida
October 2, 1987

CONTENTS

INTRODUCTION: WORLD REGIONAL GEOGRAPHY -- PHYSICAL AND HUMAN FOUNDATIONS

OJECTIVES OF THIS CHAPTER

The Introduction is an overview of world regional geography. Proceeding from a discussion of basic concepts of regionalism, scale, culture, and landscape, the natural environment, and global population patterns are explored. Additional human-geographic variables are introduced, notably political and economic forces (the latter are explored in some depth in the Opening Essay that treats global core-periphery relationships), and provide the basis for establishing the framework of world geographic realms to be pursued in this book. Following capsule vignettes on each realm, the role of regional studies in contemporary geography is reviewed. A final box introduces you to map reading and interpretation.

Having learned this basic background to the subject, you should be able to:

1. Differentiate among the various types of regions, and understand the nature of regional structure.

2. Read and interpret the maps that accompany the text, knowing such map properties as scale, projection, orientation, and point, line, and area symbols.

3. Understand the notion of culture and how it is expressed to form human landscapes in time and space.

4. Understand the general aspects of the natural environment and its evolution over the past 3 million years.

5. Explain broad, world-scale, patterns of terrain, hydrography (water occurrence), climate, vegetation, and soils, with some emphasis on the interrelationship of these environmental systems within the Koppen-Geiger climatic classification scheme.

6. See the role and importance of generalization in geography as exemplified in the use of spatial models.

7. Identify and discuss the global distribution of humanity, with emphasis on the largest population concentrations.

8. Define urbanization and be familiar with its rapidly changing dimensions across the world.

9. Appreciate how spatial expressions of politics and economics underpin the present world split into developed and underdeveloped portions.

10. Understand the importance of core areas and how, at the global scale, the developed countries function as a core that dominates the underdeveloped countries which constitute a "have-not" periphery.

11. Name, map, and differentiate among the 13 world geographic realms that comprise the spatial framework to be followed in the remaining chapters of the book.

12. Understand the broad outlines of the discipline of geography: its traditions, contemporary concerns, applications, and contributions to advancing international knowledge.

GLOSSARY

Region (1-2)

An area of the earth's surface marked by certain properties. A commonly used term and a geographic concept of central importance.

Location (2)

The position of a place or region on the earth's surface. Absolute location is that position expressed in degrees, minutes, and seconds of latitude and longitude; relative location is that position relative to the position of other places and regions. Thus, Palmyra, New Jersey, is located at 40° north latitude, 75° west longitude; relatively, this suburban community is located on the eastern bank of the Delaware River facing Tacony, a factory neighborhood in the lower northeast portion of the city of Philadelphia.

Spatial (2)

Pertaining to space on the earth's surface; synonym for geographic(al).

System (3)

Any group of objects or institutions and their mutual interactions; geography treats systems that are expressed spatially--such as regions.

Hinterland (3)

The surrounding, tributary region of a city. This area is both served by the urban center and comes under its cultural and economic influences.

Formal region (3--Box)

A type of region marked by a certain degree of homogeneity in one or more phenomena; also called uniform region or homogeneous region.

Functional region (3--box)

A region marked less by its sameness than its dynamic internal structure; because of its usual focus on a central node, also called nodal or focal region.

Culture realm (4)

A cluster of regions in which related culture systems prevail. In North America, the United States and Canada form a culture realm, but Mexico belongs to a different one.

Scale (4)

Representation of a real-world phenomenon at a certain level of reduction or generalization. In cartography, the ratio of map distance to ground distance; shown on map as bar graph, representative fraction, and/or verbal statement.

Culture (5-7)

According to Ann Larimore et al., a way of life which members of a group learn, live by, and pass on to future generations. Of course, this is but a representative definition because no single one does the concept full justice. For a sampling of other definitions, see the box on p. 6 in the text.

Cultural landscape (7-8)

The forms and artifacts sequentially placed on the physical landscape by the activities of various human occupants. By this progressive imprinting of the human presence, the physical landscape is modified into the cultural landscape, forming an interacting unity between the two. May also be regarded as a composite of artificial spaces that serves as background for the collective human existence.

Process (7)

Causal force that shapes spatial pattern or structure as it unfolds over time.

Plate tectonics (9)

Bonded portions of the earth's mantle and crust, averaging 60 miles (100 kilometers) in thickness. More than a dozen such plates exist, most of continental proportions, and they are in motion. Where they meet one slides under the other, crumpling the surface crust and producing significant volcanic and seismic (earthquake) activity. A major mountain-building force.

Pleistocene Epoch (9-12)

Recent period of geologic time that spans the rise of humanity (beginning about 3 million years ago), marked by repeated advances of continental icesheets and milder interglaciations or ice withdrawals. Although the last 10,000 years are known as the Recent Epoch, Pleistocenelike conditions seem to be continuing and the present is likely to be another Pleistocene interglaciation; the glaciers will return.

Hydrosphere (13)

The earth's surface cover of water in the form of a vast world ocean (covering just over 70 percent of the surface), frozen polar and high-mountain icesheets, and a moisture-laden atmosphere. The hydrosphere is maintained by the hydrologic cycle which circulates water in a constant repetitious cycle of evaporation, condensation, precipitation, and runoff.

Evapotranspiration (16--box)

The loss of moisture to the atmosphere through the combined processes of evaporation from the soil and transpiration by plants.

Climate (16-18)

A term used to convey a generalization of all the recorded weather observations over time at a certain place or in a given area. It represents an "average" of all the weather that has occurred there. In general, a tropical location such as the Amazon Basin has a much less variable climate than areas located, say, midway between the equator and the pole. In low-lying tropical areas the weather (the condition of the atmosphere at any given moment in time) changes little, and so the climate is rather like the weather on any given day. But, in the middle latitudes, there may be summer days to rival those in the tropics, winter days so cold that they resemble polar conditions.

Climax vegetation (18)

The final, stable vegetation that has developed at the end of a succession under a particular set of environmental conditions. This vegetative community is in dynamic equilibrium with its environment, including other ecosystems such as its animal life.

Model (20)

An idealized representation of reality created to demonstrate certain of its properties. A spatial model focuses on a geographical dimension of the real world.

Hypothetical Continent (20-21)

An idealized imaginary continent that resembles the distribution of the land surface observed at various latitudes. On this theoretical surface, the distribution of such environmental variables as climate may be mapped in a generalized way. As the Model Box on pp. 20-21 points out, this device may also be used to model the distribution of precipitation, vegetation, soil--and even population.

Comprehensive Soil Classification System (21-23)

The so-called "Seventh Approximation" or seventh attempt to classify the highly complex distribution of the world's soil. The major orders are briefly reviewed in the text, placed in their proper environmental context with respect to climate and vegetation, and mapped on pp. 24-25.

Cartogram (26)

A specially transformed map, such as Fig. I-15 on pp. 28-29, which displays the world's national populations in population-space--i.e., each country is magnified or shrunk according to its population size rather than its territorial size in "normal" space.

Urbanization (31-32)

A term with several connotations. The proportion of a country's population living in urban (metropolitan) areas is its level of urbanization. The process of urbanization involves the movement to, and clutering of, people in towns and cities, a major force in every geographic realm today. Another kind of urbanization occurs when a sprawling city absorbs rural countryside and transforms it into suburbs; in the case of the Third World city, this also generates peripheral squatter settlements.

Microstate (32-33)

The world's smallest countries in territorial and population size. Tiniest of all is the principality of Monaco on the French Riviera, whose entire territory comprises three-fourths of a square mile (less than 2 square kilometers); other European microstates are Liechtenstein, Andorra, and Vatican City. Sometimes, small political entities contain large productive populations--the Asian beehive states of Hong Kong and Singapore are notable examples, as are the federal districts of such capital cities as Washington, D.C. and Canberra, Australia.

Agriculture (34)

The purposeful tending of crops and livestock. Commercial agriculture involves the production of food surpluses to be sold for profit, almost always utilizing advanced technology; subsistence agriculture entails the production of only the necessities to sustain life in a local population, utilizing traditional (and inefficient) farming methods.

Development measures (37--box)

As listed in the box on p. 37, the leading indicators are: national product per person, labor-force occupational structure, productivity per worker, per-capita energy consumption, transportation/communication facilities per person, manufactured-metal consumption per person, and such rates as literacy, per-capita savings, and individual caloric intake.

Geographic realm (38, 41)

A major world region defined by broad multiple criteria that involve and integrate its physical, cultural, economic, political, and historical geography.

Core-Periphery Relationships (39-40)

The Opening Essay elaborates on the dependency of the Third World countries on the developed countries in the global economy. The DCs constitute the world core area--the dynamic foci of human activity that together function as the leading regions of control and change. The UDCs form the global periphery, a vast assemblage of "have-not" regions locked within a global economic system over which they have no control. At the national scale, too, UDCs suffer gross regional imbalances as small, core-based elites tightly control the underprivileged majority that inhabits the usually impoverished periphery.

Systematic geography (49)

The topical subfields of the discipline. The alternative approach to world geography: instead of using regions, individual topical aspects of geography are studied successively across the entire globe (e.g., political geography, cultural geography, and so forth).

Map projection (52)

A geometric framework within which the 3-dimensional globe is "projected" onto a flat, 2-dimensional mapping surface. Since a sphere is an undevelopable surface, at least one of the following map properties is sacrificed in the process: distance, direction, shape, or area.

Map symbols (52-53)

The point, line, and area representations that are used to portray various spatial data and distributions on map. The box on map reading and interpretation (pp. 51-53) provides appropriate illustrations and examples.

Spatial relationships (53)

An important goal of contemporary geographic analysis is to discover meaningful logical connections that explain how and why things are located where they are on the earth's surface. The Model Boxes in Chapters 1, 3, 6, and 7 show such process-response relationships in some detail; here, the historical relationship between contaminated water and cholera occurrence (p. 53) is noteworthy.

SELF-TESTING QUESTIONS

Cover this side of the page with a sheet of paper. Uncover each line after you have attempted to answer the question in the left column. If necessary, refer to textbook page(s) listed at the right.

Question	Answer	Page
REGIONAL CONCEPTS	REGIONAL CONCEPTS	
What characterizes each major world realm?	A special combination of cultural, physical, historical, economic, and organizational qualities.	1
What is meant by the term region?	An area on the earth's surface marked by certain properties.	1-2
Explain how regionalization is the geographer's means of classification.	According to selected criteria, specific meanings are assigned to certain areas.	2
Why is "Midwest" a term of relative rather than absolute location?	As an absolute location, it would have to be defined in terms of latitude and longitude; relatively speaking, "Midwest" denotes the Middle (or closer) U.S. West as seen from--or related to--the eastern seaboard.	2
Define spatial.	Pertaining to space on the earth's surface; synonym for "geographic(al)."	2
How do most regional boundaries appear in the landscape?	As broad transition zones rather than razor-sharp lines.	2
How can certain regions be conceptualized as systems?	Non-uniform regions are marked by a set of integrated activities that interconnect their various parts.	3
What are some indicators that can be used to define a city's hinterland?	Road traffic flows; telephone-call patterns; newspaper distribution; television-viewing households.	3
How does a formal region differ from a functional region?	Formal regions are marked by internal homogeneity; functional regions are defined by their structuring as spatial systems.	3 (box)
CLASSIFYING REGIONS	CLASSIFYING REGIONS	
What is a hierarchy?	A rank-ordering of phenomena, with each level subordinate to the one above it and superior to the one below.	3

Question	Answer	Page
What is the three-tier scheme of ranking cultural regions?	The largest is the culture realm, a complex large area unified by common cultural traditions; culture realms are assemblages of smaller culture regions which, in turn, consist of even smaller subregions.	4
CONCEPTS OF SCALE	CONCEPTS OF SCALE	
What is scale?	Broadly, the level of generalization; specifically, the ratio of map distance to actual ground distance.	4
What is a representative fraction?	The fraction specifying the scale of map; an RF of 1:63,360 stands for 1 map inch representing 63,360 ground inches (or 1 mile).	4
CONCEPTS OF CULTURE	CONCEPTS OF CULTURE	
What is culture?	A complex idea that cannot be expressed in a few words. Review discussion on pp. 5-8, especially the definitions in the box on p. 6.	5-8
CULTURAL LANDSCAPE	CULTURAL LANDSCAPE	
How can humans leave their imprint on the land surface?	In numerous ways, because people are agents of change; their structures and artifacts progressively transform natural into cultural landscapes.	7
What exactly is the cultural landscape?	According to Carl Sauer, the composite imprints and forms superimposed on the physical landscape by human activities.	7
What is a cultural process?	A causal force (or forces) that shapes cultural patterns, and unfolds over time.	7
What is meant by sequent occupance?	The successive stages in the evolution of a region's cultural landscape.	7
What is the iconography of a region?	According to Jean Gottmann, its unique combination of cherished symbols.	7
ENVIRONMENTAL CHANGE	ENVIRONMENTAL CHANGE	
Is the earth static or dynamic?	Dynamic, because it is in constant change--albeit in a framework of time much greater than a human life span.	8

Question	Answer	Page
What are this planet's major internal structural features?	Concentric shells comprising a core, a mantle, and a thin surficial crust.	9
What is plate tectonics?	The recently proven theory that the lithosphere consists of a set of rigid plates that are in motion, producing great stresses at their boundaries.	9
PLEISTOCENE CYCLES	PLEISTOCENE CYCLES	
What is the Pleistocene?	A recent geologic epoch (3 million to 10,000 years ago) marked by continental glaciation.	10
How many episodes of glaciation marked this Ice Age?	At least 4 major advances of continental-scale icesheets occurred, perhaps more than 10.	10
HYDROGRAPHY (WATER)	HYDROGRAPHY (WATER)	
What is the hydrosphere?	The earth's cover of water contained in surface water bodies, in icesheets, and in the atmosphere.	13
What proportion of the earth is covered by water?	The world ocean covers slightly more than 70 percent; the land surface, 29 percent.	13
What are the components of the hydrologic cycle?	Evaporating moisture from water bodies, condensation, precipitation, and surface runoff.	13
Is moisture distributed evenly across the globe?	Most decidedly not; the text discussion and Fig. I-9 (pp. 14-15) cover the details of world precipitation patterns.	16
What is the difference between humus and leaching in the shaping of soils?	Humus is the dark upper soil layer rich in nutrients from decaying organic matter; leaching is the dissolving and downward transport of near-surface nutrients by percolating water in wet climates, resulting in generally infertile soils.	15
What is meant by evapotranspiration?	The combined process of evaporation plus transpiration from plants.	16 (box)
CLIMATIC REGIONS	CLIMATIC REGIONS	
What are the general features of A climates?	These humid tropical climates are high-heat and high-moisture; important subtypes are rainforest, savanna, and monsoon climates.	16-17

What are the general features of B climates?	Dryness is dominant, and they occur in lower, middle, and higher latitudes; semiarid steppe and arid desert are the major subtypes.	17
What are the general features of C climates?	Humid, temperate, mid-latitude climates; found in eastern U.S., western Europe, and elsewhere; Mediterranean climate is a major subtype.	17
What are the general features of D climates?	Humid, cold, snowy climates most prominent in the higher-latitude interiors of large landmasses; huge annual temperature ranges.	17
What are the general features of E climates?	Cold polar climates, differentiated into tundra and icecap types.	17
What are H climates?	Undifferentiated high-altitude climates of mountains at any latitude; much like the E climates.	17
What does the world regional pattern of climates look like?	The detailed distribution is mapped in Fig. I-10 (pp. 18-19); the generalized pattern on a Hypothetical Continent is shown in Fig. I-11 (p. 20).	17-19
VEGETATION PATTERNS	VEGETATION PATTERNS	
What does the worldwide distribution of vegetation look like, and how is it related to the global climate pattern?	This is mapped in Fig. I-12 (pp. 22-23) and discussed in the text; a very close relationship to climate exists, revealed in a comparison of Figs. I-12 and I-10.	18-21
What is meant by climax vegetation?	The plant-environment equilibrium that develops in a region when climate remains the same for thousands of years (and human impacts are minimal).	18
SOIL DISTRIBUTION	SOIL DISTRIBUTION	
Is the global distribution of soil similarly related to climate and vegetation?	For the most part yes, as the world soil map (Fig. I-13 on pp. 24-25) shows; comparisons with Figs. I-12 and I-10 show the correspondences in detail, and the world population distribution map (pp. 26-27) reveals additional relationships.	21
What is the Comprehensive Soil Classification System?	The most widely used taxonomy at the present time; details are provided in the text.	21-23

MODELS IN GEOGRAPHY	MODELS IN GEOGRAPHY	
What is a model, and how does it relate to geography?	Peter Haggett's definition is the creation of an idealized representation of reality in order to demonstrate its most important properties; models are highly useful generalizations of complex empirical (real-world) relationships.	20
What is the Hypothetical Continent?	A simplified, smoothed-out, imaginary landmass on which generalized climate zones are mapped (Fig. I-11); how would vegetation and soil regions appear if they were also mapped at this level of abstraction?	20-21
SPACE AND POPULATION	**SPACE AND POPULATION**	
What does the global distribution of humanity look like?	See Figs. I-14 (pp. 26-27) and I-15 (pp. 28-29) for a pair of views.	24-31
How large is the population of the world, and what are the immediate prospects for growth?	At the beginning of 1988 the total was just under 5.1 billion; at present growth rates, by the turn of the century there will be more than 6.1 billion humans on earth.	26
Where are the four largest population concentrations located?	In descending size, East Asia, South Asia, Europe, and eastern North America.	26
What is a cartogram, and how does it help in depicting the world population distribution?	A specially transformed map based on area data rather than distance scale; accordingly, the "blowing up" and "shrinking down" of the world's countries in population-space (Fig. I-15 on pp. 28-29) enables one to see far more clearly the relative population sizes of individual countries.	26-29
What are the main features of the East Asian population cluster?	China's Huang He (Yellow) and Chang Jiang (Yangtze) river valleys; dominantly rural population.	27-28
What are the main features of the South Asian cluster?	Pakistan's Indus Valley, India's Ganges lowland, and all of Bangladesh in the Ganges Delta; heavily rural.	28-30
What are the main features of the European cluster?	A central east-west population axis, oriented to industrial resources and highly urbanized.	30
What are the main features of the North American cluster?	A European-type pattern, dominated by Megalopolis along the northeastern U.S. seaboard.	30

Question	Answer	Page
Where are other notable population clusters located?	Southeast Asia (especially Indonesia [Jawa] and the deltas of Vietnam), West Africa, and Egypt's lower Nile Valley.	31
URBANIZATION	URBANIZATION	
What are the chief characteristics of <u>urbanization</u>?	A process of population concentration involving the growth of cities as well as the in-migration of new residents from rural areas.	31
What is the chief worldwide urban trend in the 1980s?	Universal rapid growth in the less developed continents, stability in the advanced countries; details provided in the graph and table on pp. 31 and 32.	31-32
POLITICAL GEOGRAPHY	POLITICAL GEOGRAPHY	
How many political units are there today?	In the mid 1980s, about 160 national states and 50 dependent territories.	32
Where are some current pressure zones in the world boundary framework?	Cyprus, Lebanon, Iran-Iraq, Israel, Syria are examples; since the textbook went to press, the Kashmir dispute between India and Pakistan has heated up again; new disputes will undoubtedly arise in the future.	33
ECONOMIC GEOGRAPHY	ECONOMIC GEOGRAPHY	
What is the central concern of economic geography?	The varied ways in which people earn a living, and how the goods and services they produce are spatially distributed and organized.	33-34
What is humanity's leading occupation?	Agriculture: the purposeful tending of crops and livestock.	34
What is the difference between subsistence and commercial agriculture?	Subsistence farming produces only the necessities to sustain life in a local population; commercial agriculture produces food surpluses that are sold for profit, and is almost always conducted at a far higher level of technology.	34-36

UNDERDEVELOPMENT	UNDERDEVELOPMENT	
What are DCs and UDCs, and how do they differ?	DCs and UDCs are, respectively, developed and underdeveloped countries; UDCs are the "have nots," disadvantaged Third World countries whose economies rank far behind those of the developed world.	36-37
How is economic development measured?	According to a number of indices that facilitate international comparison; the box on p. 37 lists several of the most commonly used variables.	37-38
Are UDCs internally homogeneous in their low level of economic development?	No: nearly every UDC is characterized by geographic imbalances in its regional structuring; the capital might be bustling port and commercial center, but its outskirts are usually squalid shacktowns (see photo on p. 326) and the rural areas beyond are mired in abject poverty.	38

GEOGRAPHIC REALMS	GEOGRAPHIC REALMS	
What is a geographic realm, and how does it differ from a culture realm?	Whereas a culture realm is a large region based exclusively on cultural criteria, a geographic realm is based on broader multiple criteria--reflecting physical, economic, political, urban, historical, and population as well as cultural geography.	41
What does the distribution of world geographic realms look like?	This framework, which underlies the organization of the rest of this book, is mapped on pp. 42-43 (Fig. I-19) and should be studied carefully.	42-43
What are the names of the 13 geographic realms?	The 5 developed realms (Chapters 1-3, including two vignettes) are Europe, Australia/New Zealand, the Soviet Union, North America, and Japan; the 8 underdeveloped realms are Middle America, South America, Southwest Asia North Africa, Subsaharan Africa, South Asia, The Chinese World, Southeast Asia, and the Pacific.	41-48
What are the major characteristics of each realm?	They are too numerous to list here; study pp. 41-48 closely--the matching question in this chapter of the Study Guide will also be helpful to you.	41-48

REGIONAL GEOGRAPHY	REGIONAL GEOGRAPHY	
What does the word "geography" mean specifically?	Literally, it means to write about the earth.	48
How old is the practice of regional geography?	As old as the discipline itself, which is over 3,000 years old and is often called the "parent" of the sciences.	48
Besides the regional or area-studies approach, what are the other major traditions of geography?	These are: (1) the earth science tradition, focusing on physical geography; (2) the culture-environment tradition, involving human-habitat interaction; and (3) the spatial organization tradition, treating the locational arrangement of spatial distributions.	48
Which tradition is dominant in contemporary geography?	The spatial organization tradition holds center stage, but regional studies appear to be making a comeback.	49
What are the systematic subfields of geography?	The topical branches of the discipline that cut across the world realms, such as political or economic geography; ten systematic subfields will be reviewed in this book, one at the outset of each subsequent chapter--the scheme is diagramed in Fig. I-22 on p. 50.	49

MAP INTERPRETATION	MAP INTERPRETATION	
Define cartography.	The essence of this geographical method is covered on pp. 51-53. The term itself is defined in the Glossary, a further instructional aid in this textbook that you should be aware of.	51-53
How is map scale expressed?	By the representative fraction and the bar graph.	4/51
How can one discern orientation on a map?	By referring to the gridded framework of latitude and longitude, designators of absolute location.	51-52
What is a map projection?	The geometric latitude-longitude framework of a map, which is by definition distorted because the grid of a 3-dimensional sphere cannot be transferred to a 2-dimensional flat map without sacrificing certain map properties.	52
What are the different classes of map symbols?	Point, line, and area symbols as shown in Figs. 1-23 and I-24; there are also volume symbols, which are not shown.	52-53

MAP EXERCISES

MAP COMPARISON

1. Comparing Figs. I-5 (pp. 10-11)) and I-6 (p. 12), what generalizations can be made about the landforms that are located along plate boundaries, especially where plates are pushing against one another?

2. Comparing Figs. I-5 (pp. 10-11) and I-8 (p. 13), what generalizations can be made about the location of Pleistocene icesheets in the Northern Hemisphere and the distribution of mountain ranges?

3. Comparing Figs. I-14 (pp. 26-27) and 5-3 (pp. 312-313), what kinds of agricultural activities are associated with the densest concentrations of humanity? (These activities are briefly outlined on pp. 312-314 of the text).

4. Comparing Figs. I-19 (pp. 42-43) and 6-13 (pp. 378-379), what generalizations can be made--continent by continent--concerning the distributions of oil-(petroleum-) producing countries and the developed realms?

5. What generalizations about the eastern half of the United States may be made regarding interrelationships between climate, soil, and vegetation? (Compare Figs. I-10 through I-13, respectively, found on pp. 18-25).

6. On Fig. I-14 (pp. 26-27), compare and contrast the internal patterns of the world's 4 largest population clusters. Can some of the variations be accounted for by differences in environmental patterns as mapped in Figs. I-5, I-10, I-12, and/or I-13?

7. On Fig. I-15 (pp. 28-29), how does the cartogram mapping technique make it easier to pick out the more heavily populated parts of the world when compared to the conventional distribution map (I-14) on pp. 26-27? Alternatively, what are some shortcomings of the cartogram as a comparative cartographic device?

8. Compare the advantages and disadvantages of generalizing world climate-region patterns (Figs. I-10 and I-11 on pp. 18-19 and 20, respectively), with special emphasis on the mainland United States.

9. Carefully compare the map of world political units with the population cartogram (Figs. I-17 and I-15 on pp. 34-35 and 28-29, respectively). List ten countries that are prominent in the cartogram but not on the world political map, and ten countries that are large in territorial size on the political map but modest in appearance on the cartogram.

10. Utilizing Figs. I-23 and I-24 (pp. 51, 52) as a guide, review all of the maps in the Introduction chapter in order to find examples of point and line symbols (area symbols, of course, abound in these maps). At a minimum, you should be able to find three examples of line symbols and one figure exhibiting point symbols.

MAP CONSTRUCTION

(Use outline maps at the end of this chapter)

1. In order to examine and study the interrelationships among environmental systems, draw in the following for each of the world's continents:

   a. in light red, sketch in all major highland areas (red, brown, and orange zones in Fig. I-5 on pp. 10-11 in the textbook)

   b. in green, sketch in the wettest zones (the color closest to the top of the legend color-box in Fig. I-9 on pp. 14-15); in yellow, sketch in the driest zones (color closest to the bottom of the color-box)

   c. in black or blue, draw in all climate-zone boundaries from Fig. I-10 (pp. 18-19) and label each climate region with its appropriate Koppen-Geiger letter symbol.

2. Using Fig. I-14 (pp. 26-27) as a guide, draw in and label each of the major and minor world population concentrations discussed on pp. 26-31 in the text.

3. Using Fig. I-19 (pp. 42-43) as a guide, draw in the boundaries of all of the 13 world geographic realms and label each regional entity; you should be able to do this by memory if you have learned Fig. I-19 first.

PRACTICE EXAMINATION

SHORT-ANSWER QUESTIONS

Multiple Choice

1. Which of the following cartographic devices when deployed on a map will tell you its scale:

   a) point symbol
   b) legend
   c) representative fraction
   d) map projection
   e) latitude

2. The islands of the Caribbean Sea belong to which of the following realms:

   a) North America
   b) Middle America
   c) South America
   d) Subsaharan Africa
   e) Europe

3. The Mediterranean climate is classified under which of the following Koppen-Geiger letters:

   a) C
   b) H
   c) R
   d) B
   e) M

True-False

1. The world's population is now almost 4 billion in total size, and is expected to be about 4.2 billion by the year 2000.

2. Southeast Asia does not rank among the world's four largest concentrations.

3. The Pleistocene Epoch ended thousands of years before the appearance of the human species.

Fill Ins

1. The dominant religion of the largest country in the realm called South Asia is __________.

2. The ______________ region, also known as the nodal or focal region, is marked not by an internal sameness but by its dynamic internal structuring.

3. Desert and steppe climates both belong to the Koppen-Geiger class included under the letter ______.

Matching Question on World Realms

_____	1.	Caste system	A.	Europe
_____	2.	Communist since 1949	B.	Australia/New Zealand
_____	3.	World's largest country in area	C.	Soviet Union
_____	4.	Caribbean islands	D.	North America
_____	5.	Moslem culture	E.	Japan
_____	6.	Transformer of modern world	F.	Middle America
_____	7.	Huge variety of "community" religions	G.	South America
_____	8.	Indochina	H.	Southwest Asia/North Africa
_____	9.	Melanesia and Micronesia	I.	Subsaharan Africa
_____	10.	Two pluralistic national societies	J.	South Asia
_____	11.	Victim of nuclear attack	K.	The Chinese World
_____	12.	Dominant Iberian imprint	L.	Southeast Asia
_____	13.	Western outpost in oceanic world	M.	The Pacific World

ESSAY QUESTIONS

1. Identify and discuss the major geographic characteristics of your home region. Include, in your narrative, substantive answers to the following questions: Is it a formal or functional region, or does it contain aspects of both? How strong a role does urbanism play and, if prominent, are you located in the urban center itself or the hinterland? Are the boundaries of your region sharply defined or do they appear as wide zones of transition?

2. The cultural landscape of a region is one of its most important geographic components. Define and discuss this concept, taking into account the role of culture itself, its translation into a spatial expression on the land surface, and the significance of the special qualities that offset one cultural landscape from another.

3. Water has been shown to be one of the most significant variables in the natural environment. Discuss the broad global patterning of precipitation, and the interrelationships between that distribution and world patterns of climate, vegetation, and soils.

4. The spatial distribution of population has been called the most essential geographic expression because it represents the totality of human adjustments to the environment at that moment in time. Identify and discuss the global distribution of humankind, focusing on the four largest population concentrations and the reasons why so many people have clustered in those locations.

5. The geographic realm is the regional unit that forms the basis of the study of world geography in this book. Discuss what is meant by this regional term, and using examples chosen from the 13-part scheme introduced on pp. 41-48, list three leading features of three different realms.

TERM PAPER POINTERS

In this section of the Study Guide, term paper topics will be discussed and some tips provided. Research paper topics can readily be selected from among the major themes emphasized in each chapter, and it should be easy for you to identify several in the Introduction. However, since the aim of this opening chapter is to provide a broad background on world physical and human geography prior to a realm-by-realm survey, it is presumed that your term papers will deal with topics rooted in the coming regional chapters. Nonetheless, you will want to keep in mind the topical coverage of the Introduction chapter because of its worldwide scope as well as the introduction of ideas and concepts that are developed further later in the book. For example, if you decided to do a term paper comparing the traditional agricultural systems of India and China, there is much valuable background material here in the Introduction; among the topics relevant to such a study are environmental differences (apparent in world maps of climate, precipitation, landscapes, and soils) and the economic-geographical properties of densely-populated UDCs, and among the conceptual themes first introduced here are cultural landscapes, geographic realms, economic development, various spatial relationships, and agriculture itself.

One of the most important tasks in preparing a research paper is finding the appropriate literary sources. The chapter-end bibliographies have been revised and updated with exactly that in mind, and are well worth consulting after you have chosen your term-paper topic. At the same time, it is a good idea to become acquainted with the professional journals of geography that are held by your library. Most of the topics treated in this textbook are covered in these periodicals, which are indexed in the appropriate reference catalogues (and annually in the journals themselves). For the uninitiated researcher, The Geographical Review, The Professional Geographer, the Journal of Geography, and the Journal of Cultural Geography are most convenient to start with. At the more advanced level, there is the Annals of the Association of American Geographers, the Geographical Journal, and Progress in Geography; only those with mathematical training should consult Geographical Analysis and Economic Geography (the latter, however, is quite appropriate for newcomers in volumes that were published before 1970). In larger libraries, you will also find journals on specific subdisciplines (e.g., The Journal of Historical Geography, Urban Geography, Physical Geography) as well as certain world regions (e.g., Journal of Tropical Geography, Soviet Geography: Review and Translation, Caribbean Geography). In the interdisciplinary arena, geography is frequently treated in Landscape, Environment and Behavior, and Economic Development and Cultural Change. Moreover, regional geography topics are widely covered in the regional studies journals that are multidisciplinary in nature; some of the leading ones are the Journal of Asian Studies, the Latin American Research Review, and the Journal of Modern African Studies--do not overlook the journal(s) that covers the part of the world you are studying. It is also worthwhile to remember the bibliographical footnote on p. 131 in the textbook, which steers you toward book and serial publications on individual countries. Finally, there are several bibliographies available (some are referenced in the chapter-end listings of further readings); the most significant is Current Geographical Publications which is published at regular intervals by the American Geographical Society, an updating of its central catalogue that many university libraries also possess in the form of a series of hardbound volumes in their reference department.

In subsequent Study Guide chapters, this section on term paper topics and pointers will serve two purposes: it will offer suggestions about the realm covered in the book chapter and the systematic subfield of geography treated in the Systematic Essay.

SPECIAL EXERCISE

Each chapter in this Study Guide concludes with a special exercise, designed to provide an enrichment experience by building upon a major idea introduced in the parallel book chapter. Model Boxes afford an especially good opportunity to do this, and we focus here on the Hypothetical Continent as an expression of modeling in geography.

Model Box 1 (text pp. 20-21) introduces the notions of generalization and abstraction in geography. The Hypothetical Continent (p. 20) is presented as a model whose simplification of the complex pattern of real-world climates is readily apparent in a careful comparison of Figs. I-10 (pp. 18-19) and I-11 (p. 20). Using tracing paper, make two copies of the outline of the Hypothetical Continent leaving the land area blank. Then, by referring to Figs. I-12 (pp. 22-23) and I-13 (pp. 24-25), prepare generalized vegetation and soil region maps that fit onto the Hypothetical Continent in the same manner as the smoothed-out climate region map on p. 20. Finally, make observations about the similarities and differences among the three "model" distributions, basing your comments on what you have learned about the interrelationships of climate, soil, and vegetation (review text pp. 12-24).

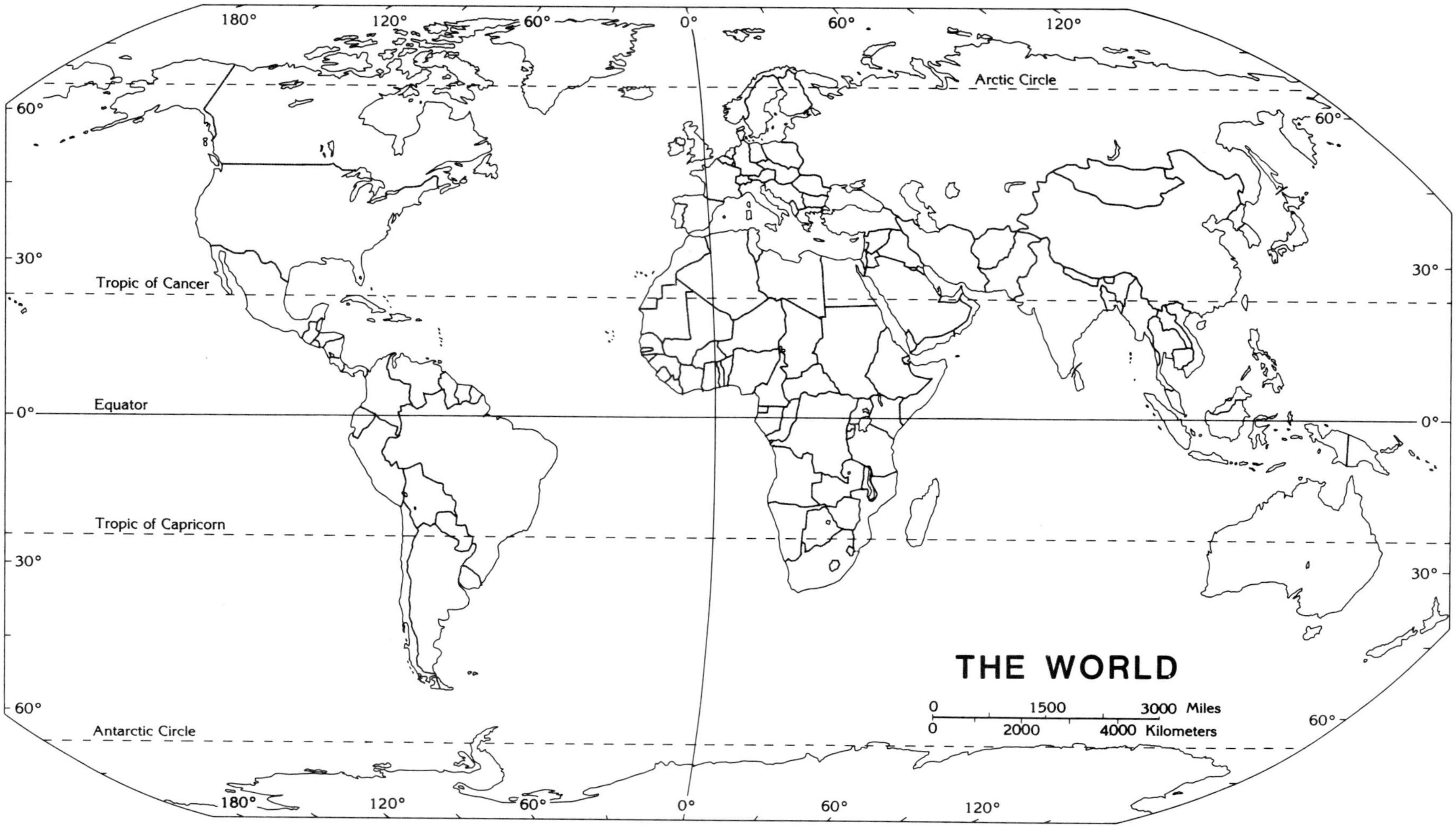
THE WORLD
0 1500 3000 Miles
0 2000 4000 Kilometers
Arctic Circle
Tropic of Cancer
Equator
Tropic of Capricorn
Antarctic Circle
180°
120°
60°
0°
60°
120°
30°
0°
30°
60°

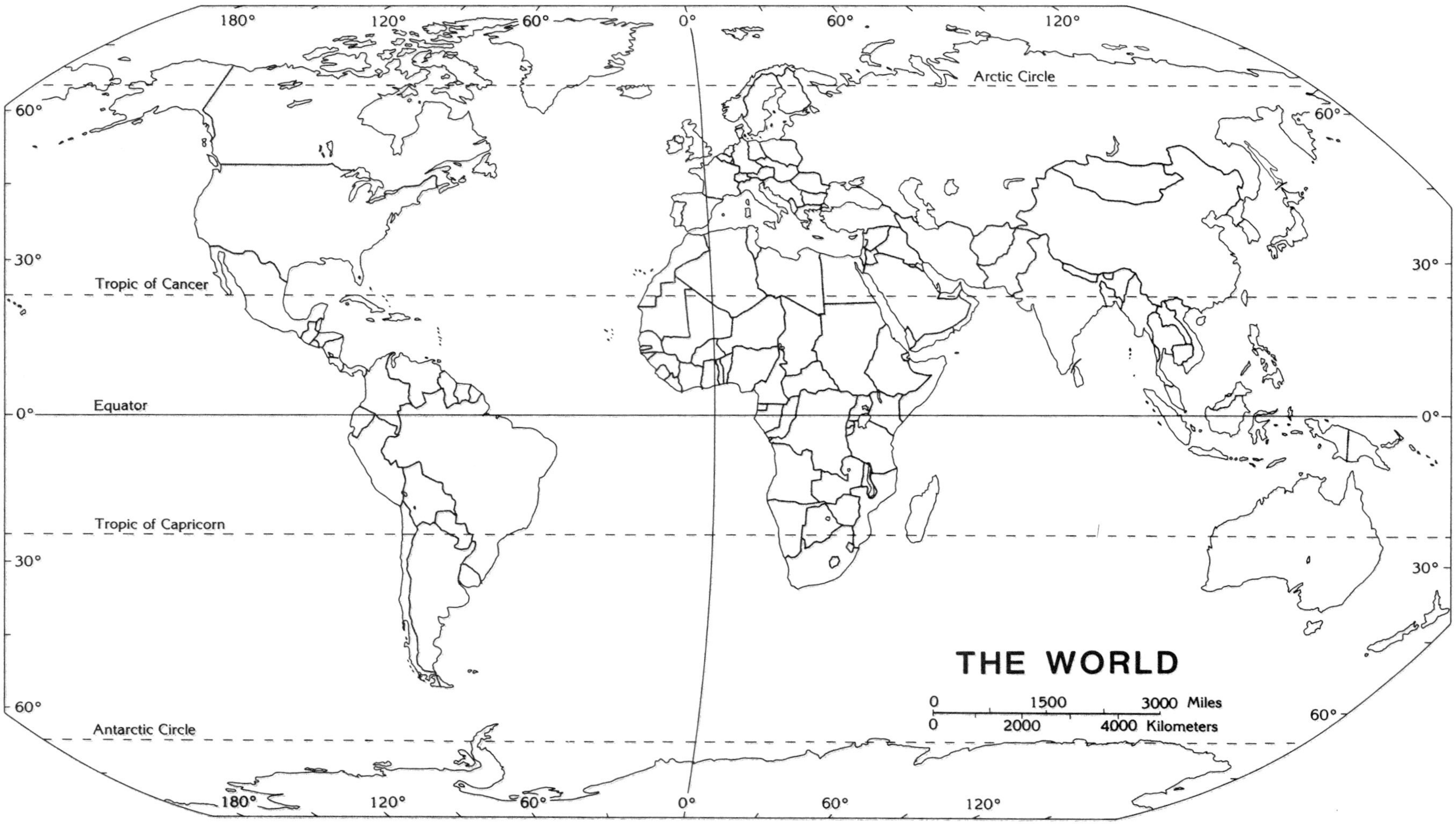
THE WORLD
0 1500 3000 Miles
0 2000 4000 Kilometers
Arctic Circle
Tropic of Cancer
Equator
Tropic of Capricorn
Antarctic Circle
180°
120°
60°
0°
60°
120°
30°
0°
30°
60°

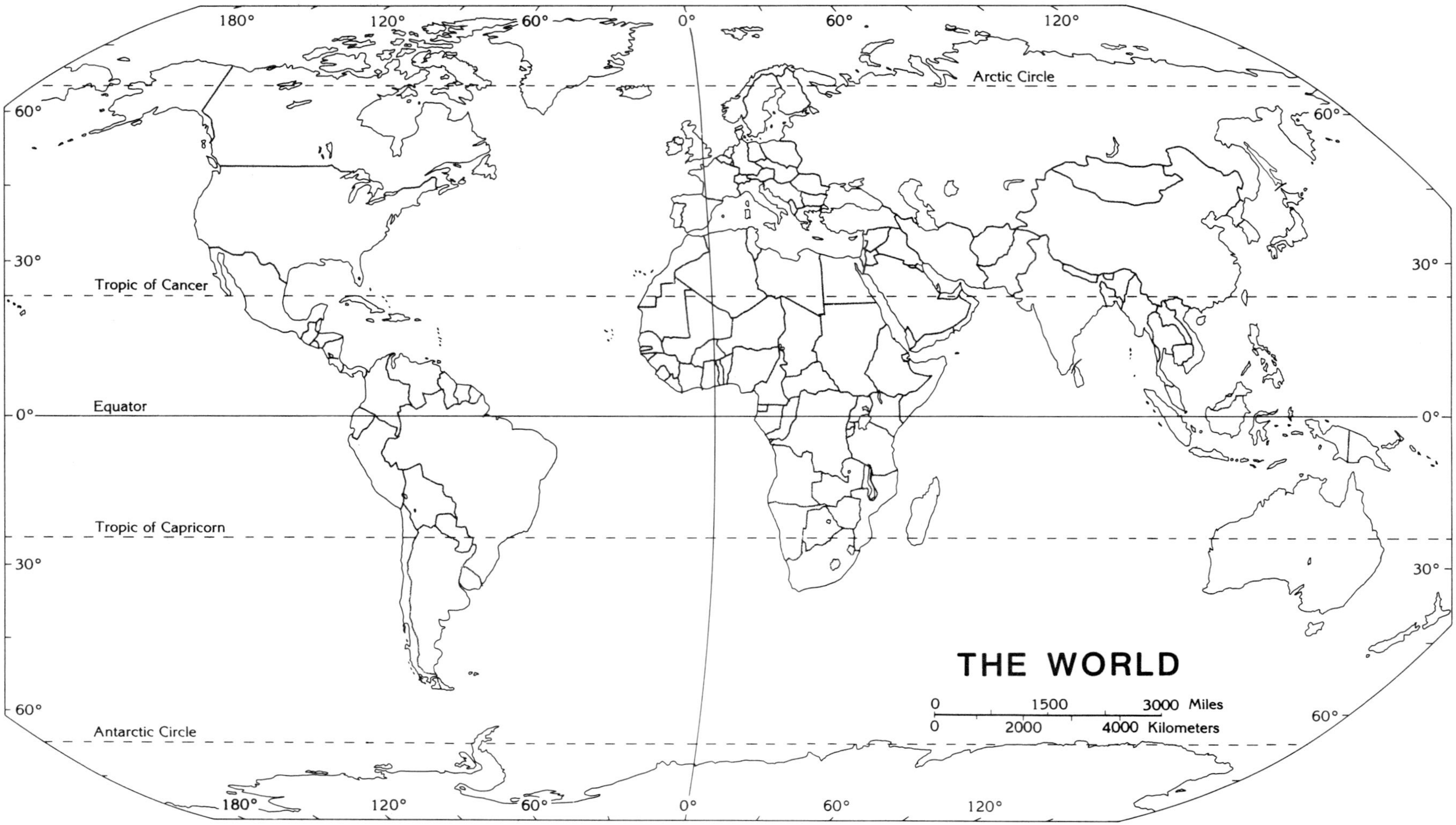
THE WORLD
0 1500 3000 Miles
0 2000 4000 Kilometers
Arctic Circle
Tropic of Cancer
Equator
Tropic of Capricorn
Antarctic Circle
180°
120°
60°
0°
60°
120°
30°
0°
30°
60°

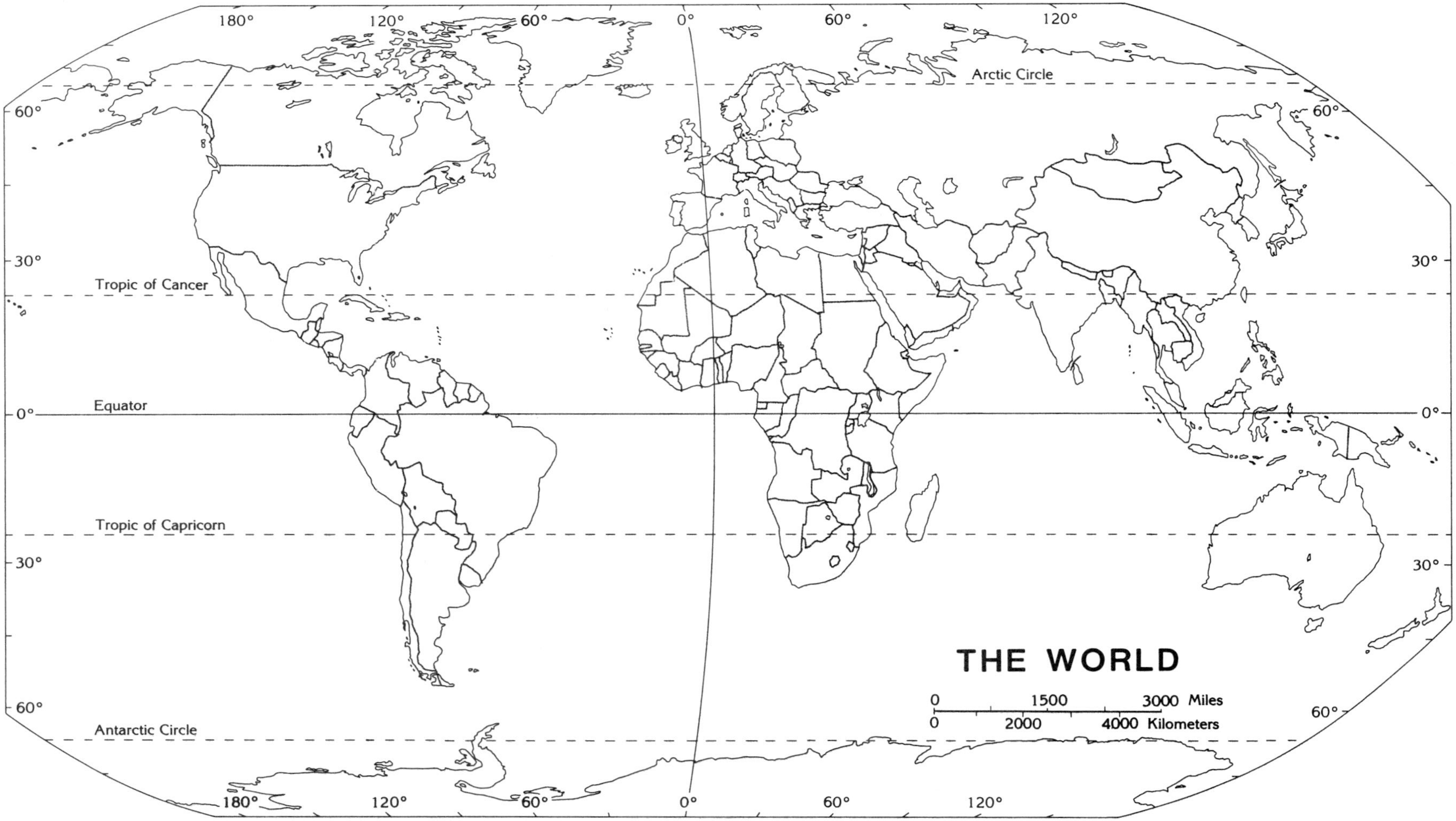
THE WORLD
0 1500 3000 Miles
0 2000 4000 Kilometers
Arctic Circle
Tropic of Cancer
Equator
Tropic of Capricorn
Antarctic Circle
180°
120°
60°
0°
60°
120°
30°
0°
30°
60°

CHAPTER 1: RESILIENT EUROPE—CONFRONTING NEW CHALLENGES

OBJECTIVES OF THIS CHAPTER

Chapter 1, which is the first regional chapter as well as the first to treat the developed-world realms, focuses on the geography of Europe. It begins with the first Systematic Essay--population geography--which explores both this topical subfield of geography as well as its application to the modern European scene. Following an introduction to Europe's advantageous geographical features, its physical, historical, economic, political, and urban geography are surveyed in their realm-wide contexts. The rest of the chapter is devoted to an examination of Europe's internal regions, proceeding through the Bristish Isles, Western, Northern, Mediterranean, and Eastern Europe. The concluding section treats European supranationalism and ponders the prospects for further unification.

Having learned the regional geography of Europe, you should be able to:

1. Understand the elements of population geography, especially the Demographic Transition Model.
2. Identify Europe's remarkable geographic properties that propelled this modestly-sized corner of the world into international prominence and advanced economic development.
3. Understand Europe's broad physical geography, especially its subdivisioning into four major physiographic regions.
4. Appreciate the rich historical heritage of European society, whose modern foundation rests upon numerous economic and political revolutions.
5. Discern the current geographical dimensions of the realm, including the problems faced by overindustrialized areas in a new postindustrial age.
6. Explain the basic workings of the Von Thunen model, and its empirical application to 20th-century Europe.
7. Name the various major internal regions of Europe, and the countries they contain.
8. Understand such basic politico-geographical concepts as nation-state, devolution, irredentism, the organic theory of state evolution, and supranationalism.
9. Identify the various subregions of the British Isles as well as Western, Nordic, Mediterranean, and Eastern Europe, and be familiar with the major geographic qualities of each of them.
10. Locate the major features of Europe on a map, including its countries, prominent physical regions, primary rivers, leading industrial areas, and largest urban centers.

GLOSSARY

Pattern (60)

The geometric arrangement of a spatial distribution.

Dispersion (60)

The extent of spread of a spatial distribution.

Density (60)

The frequency of occurrence of a phenomenon within a given area.

Demographic Transition (61-62)

Three-stage model, based on the United Kingdom's experience, of changes in population growth exhibited by countries undergoing modern industrialization. High birth and death rates are followed by plunging death rates, producing a huge net population gain; this is followed by the convergence of birth and death rates at a low overall level.

Migration (62-63)

A change in residence intended to be permanent. There are many different categories of migration such as voluntary, involuntary, internal, international, and rural to urban.

Lingua franca (69)

The term derives from Frankish language, and applies to a tongue spoken in ancient Mediterranean ports that consisted of a mixture of Italian, French, Greek, Spanish, and even some Arabic. Today, it refers to a "common language," a second language that can be spoken and understood by many peoples, although they speak other languages at home.

Mercantilism (72)

Protectionist policy of European states during the sixteenth to the eighteenth centuries that promoted a state's economic position in the contest with other countries. The acquisition of gold and silver and the maintenance of a favorable trade balance (more exports than imports) were central to the policy.

Nation-state (72-box)

A country whose population possesses a substantial degree of cultural homogeneity and unity.

Industrial Revolution (73-74)

A major turning point in a region's economic development where manufacturing becomes a leading growth activity, engendering further technological breakthroughs, capital investments, urbanization, and (eventually) a demographic transition. Early 19th century Western Europe and the northeastern United States first underwent this transformation, which comprises a number of steps, and subsequent spreading took place throughout the rest of the developed world. Certain emerging countries in the Third World are approaching this "takeoff" stage, but most are not as we shall see in Chapters 4-10.

Von Thunen Model (76-78)

Explains the location of agricultural activities in a commercial, profit-making economy. A process of spatial competition allocates various farming activities into concentric rings around a central marketplace, with profit-earning capability the determining force in how far a crop locates from the market. The original (1826) Isolated State model now applies to the continental scale (Fig. 1-12B on p. 77); in certain Third World areas, however, transportation conditions still produce an application reminiscent of the original (Fig. 6-30 on p. 409).

Agglomeration (78)

Process involving the clustering of people or activities. Often refers to manufacturing plants and businesses that benefit from close proximity because they share skilled-labor pools and technological and financial amenities.

Organic theory (79)

Concept that suggests that the state is in some ways analogous to a biological organism, with a life cycle that can be sustained through cultural and territorial expansion.

Geopolitik (Geopolitics) (79)

A school of political geography that involved the use of quasi-academic research to encourage a national policy of expansionism and imperialism. Its origins are attributed to a Swede by the name of Kjellen, but its most famous practitioner was a German named Karl Haushofer.

Spatial interaction (80)

Movement of any sort across geographic space, usually involving the contact of people in two or more places for the purposes of exchanging goods or ideas.

Complementarity (80)

Regional complementarity exists when two regions, through an exchange of raw materials and finished products, can specifically satisfy each other's demands.

Intervening opportunity (80)

The presence of a nearer opportunity or supply source that diminishes the attractiveness of sites farther away.

Transferability (80)

The capacity to move a good from one place to another at a bearable cost; the ease with which a commodity may be transported.

Primate city (83)

A country's largest city--ranking atop the urban hierarchy--most expressive of the national culture and usually (but not always) the capital city as well.

Metropolis (84)

An urban complex consisting of the core or central city and the surrounding ring of suburbs, increasingly an outer city in its own right in the most developed countries.

Central business district (CBD) (84)

The heart of a city, the CBD, is marked by high land values, a concentration of business and commerce, and the clustering of the tallest buildings.

Conurbation (89)

General term used to identify large multi-metropolitan complexes formed by the coalescence of two or more major urban areas. The Boston-Washington Megalopolis along the U.S. northeastern seaboard is an outstanding example.

Devolution (92-box)

In political geography, the disintegration of the nation-state as a result of emerging or reviving regionalism.

Break-of-bulk point (95)

A location along a transport route where goods must be transferred from one carrier to another. In a port, the cargoes of oceangoing ships are unloaded and put on trains, trucks, or perhaps smaller boats for inland distribution.

Site (96)

The internal locational attributes of a place, including its local spatial organization and physical setting.

Situation (96)

The external locational attributes of a place; its relative location or position with reference to other non-local places.

Transhumance (104)

Seasonal movement of people and their livestock in search of pastures. This movement may be vertical--that is, into highlands during the summer and back to lowlands in winter--or horizontal, in pursuit of seasonal rainfall.

Acid rain (107-108 box)

Highly corrosive solution on fragile ecological systems, which originates when the air pollutants sulfur dioxide and nitrogen oxide combine with water vapor prior to precipitation. Devastating environmental effects such as deforestation and fish kills; particularly vulnerable areas are wilderness zones downwind from industrial centers.

Irredentism (119)

A policy of cultural extension and potential political expansion aimed at a national group living in a neighboring country.

Balkanization (120-box)

The fragmentation of a region into smaller, often hostile, political units.

Supranational (126)

A venture involving three or more national states--political, economic, or cultural cooperation to promote shared objectives. The European Common Market is such an organization, as is OPEC.

European Economic Community (127)

Now called simply the European Community (Common Market), it currently consists of twelve countries--France, Italy, West Germany, Belgium, the Netherlands, Luxembourg, the United Kingdom, Ireland, Denmark, Greece, Spain, and Portugal.

European Free Trade Association (127-128)

The so-called "Outer Five"--Sweden, Norway, Switzerland, Austria, and Iceland. Formerly a counterpart to the Common Market, EFTA was weakened by the departure of Britain and Denmark to join the EC in 1972; Finland, preserving its strict neutrality status, maintains an associate membership in EFTA.

COMECON (129)

The Council for Mutual Economic Assistance, formed by the Soviet Union which ordered all of its Eastern European satellites to join (Yugoslavia and Albania managed to resist). On paper a socialist alternative to the European Community, COMECON is merely another device whereby the U.S.S.R. enforces its hegemony over the region.

SELF-TESTING QUESTIONS

Cover this side of the page with a sheet of paper. Uncover each line after you have attempted to answer the question in the left column. If necessary, refer to the textbook page(s) listed at the right.

Question	Answer	Page
POPULATION GEOGRAPHY	POPULATION GEOGRAPHY	
How does population geography differ from demography?	Demography studies population from an interdisciplinary perspective; population geography treats the distribution, composition, growth, and movement of people related to spatial variations in living conditions on the surface of the earth.	60
Why is population distribution one of geography's most essential expressions?	Because the way in which people have arranged themselves in geographic space represents the totality of the adjustments they have made to their environment over time.	60
What is meant by the term _ecumene_?	The habitable areas of a region or the world where people have settled.	60
Define _density_, _dispersion_, and _pattern_, which are all properties of population distribution.	Density is the frequency of occurrence of a phenomenon in a region; dispersion is its extent of spread; pattern is its geometric arrangement.	60
What is the difference between _arithmetic_ and _physiologic_ density?	Arithmetic density is the number of people per unit area; physiologic density is the number of people per unit area of arable or cultivable land.	60
What is the Demographic Transition Model?	A generalization of the stages of demographic change experienced by countries undergoing Industrial Revolutions.	61-62
Define _migration_.	The purposeful relocation of one's permanent residence.	62
EUROPE'S CHARACTERISTICS	EUROPE'S CHARACTERISTICS	
What are Europe's chief geographical resources?	Its unmatched advantages of scale and proximity, a diverse environment and cultural mosaic, facilitating innovation and easy contact with the rest of the world.	59, 63-64

Where is Europe's eastern boundary?	Some scholars insist it is the Ural Mountains, while others argue that all of western Russia is a transition zone; the text uses the boundary between Eastern Europe and the Soviet Union.	67-box
PHYSICAL LANDSCAPES	PHYSICAL LANDSCAPES	
What are the major characteristics of the Central Uplands?	Forms a transition zone between the Alps and the North European Lowland; a resource-laden belt where Industrial Revolutions and cities emerged in the 19th century.	65-66
What are the major characteristics of the Alpine Mountains?	The Alps and their outliers (the Pyrenees, Dinaric Alps, and Carpathians) form a high-mountain backbone, separating Mediterranean Europe from the rest of the realm-- but have not been a serious obstacle to trade and communication.	66
What are the major characteristics of the Western Uplands?	Rugged, older highlands often located along stormy oceanic fringes; glaciated in the north, a further hindrance to large-scale settlement.	66
What are the major characteristics of the North European Lowland?	Densest populations of the realm are found here; historic route of human contact and migration, yet much internal variation in generally low-lying terrain; also contains much of Europe's most productive agriculture.	66-67
HERITAGE OF ORDER	HERITAGE OF ORDER	
What area was the heart of the ancient civilization of Greece? And how did the Greeks politically organize their territory?	The eastern Mediterranean formed the core of a large empire, that was eventually defeated by Rome. Greece was organized into leagues of city-states; internal squabbling, especially between Athens and Sparta, hastened the Greek downfall.	68
What additional contributions did the Romans make to empire-building?	Politico-territorial organization on a much wider scale, extending from Britain to Persia and from the Black Sea to the lower Nile Valley; internal diversity was a strength, fostering trade and exchange of ideas and innovations; unparalleled advances in agriculture, political authority, urbanization, transportation, and development of areal functional specialization.	68-70

MEDIEVAL AND RENAISSANCE EUROPE

MEDIEVAL AND RENAISSANCE EUROPE

Question	Answer	Page
What were the lasting contributions of the Romans through the Dark Ages?	Their language and its offshoots, Christianity, educational traditions.	70
What was the overriding politico-geographical trend in Europe between the fall of Rome and the Renaissance?	The fragmentation of large political units into tiny feudal territories; some empires emerged, as did (eventually) elements of the realm's modern political map.	70-71
What were the main trends of the Renaissance?	Reviving monarchies, early nation-state formation, beginnings of overseas colonial empires, political nationalism, and renewed interest in Greek and Roman achievements.	72
What is *mercantilism*?	The main form of economic nationalism--the acquisition of colonial territories, gold and silver, and favorable international trade--a policy actively promoted by the state.	72
What is a *nation-state*?	A political unit comprising a clearly defined territory and inhabited by a substantial population possessing binding emotional and other ties expressed through a number of institutions.	72-box

THE REVOLUTIONS

THE REVOLUTIONS

Question	Answer	Page
What was the agrarian revolution?	The transformation of European agriculture through changes in land tenure, improvements in farming techniques, equipment, storage, and distribution systems.	73
What was the Industrial Revolution?	The rapid growth of mass manufacturing through mechanization that triggered far-reaching social and economic change, demographic transition, and mass urbanization.	73-74
How did the political revolution after 1780 forever change Europe?	Monarchies gave way to republics, democracy flourished, and nationalism became the dominant political force.	74

NEW GEOGRAPHIC DIMENSIONS

NEW GEOGRAPHIC DIMENSIONS

Question	Answer	Page
What is the Von Thunen model and what are its lasting virtues?	Johann Heinrich Von Thunen in 1826 introduced his model of commercial agricultural spatial organization which still applies to contemporary Europe at the macro-scale (see text pp. 77-78). Basically, agriculture will organize itself into a series of concentric zones around the major food market(s), with the most profitable farming activity located in each ring.	75-78

Name the factors of industrial location enunciated by Alfred Weber.	"General" factors such as transport costs, and "special" factors such as perishability in the case of certain foods; "regional" factors (transport and labor costs), and "local" factors of agglomeration and deglomeration; transport costs were seen as highly critical, with plants most attracted to locate at the least-transport-cost site(s).	78-79
What was Friedrich Ratzel's contribution to political geography?	The organic theory of state evolution, which held that states passed through growth stages in which additional territory must be added; such determinism gave rise to Geopolitik (geopolitics), which helped lay the philosophical foundation for German military expansion.	79
EUROPE TODAY	EUROPE TODAY	
Is Europe an overall homogeneous regional unit?	To a surprising degree it is not--language, religion, and race are marked by much diversity; yet this diversity, through various subregional complementarities and cooperation, has been overcome to form a well-unified realm.	80
Define the three principles of spatial interaction enunciated by Edward Ullman.	Complementarity, which exists when two places, through an exchange of goods or services, can specifically satisfy each other's demands; transferability, the capacity to move an item from one place to another at a bearable cost; and intervening opportunity, the presence of a nearer trade opportunity that usurps the attractiveness of places farther away.	80
What is Benelux?	Europe's first modern attempt at economic cooperation at the international level--formed in 1944, this organization was comprised of Belgium, the Netherlands, and Luxembourg, and all three countries joined the Common Market at its founding in 1957.	82
What are some signs of current industrial decline in Britain, which may be symptomatic for Europe as a whole.	Growing industrial inefficiency, low and declining worker output, lagging productivity, increasingly obsolescent factories, declining capital investments, and destructive government-labor confrontations.	82

Question	Answer	Page
What are some warning signs about possible future agricultural decline in Europe?	Imports remain at a high level while the number of European farmers declines; mechanization is modestly-scaled and not all that widespread.	83
What is the status of Europe's transportation network?	Always highly efficient, it is among the most advanced on earth; rail and inland water transport is unrivaled anywhere, and highway and air routes are good and improving.	83
EUROPEAN URBANIZATION	EUROPEAN URBANIZATION	
Where does Europe rank among the urbanized realms?	Near the top as Table I-1 (p. 32) shows; Western Europe's countries rank in the 80-percent-and-higher range, and even lagging Eastern Europe exhibits urban percentages in the 55-65 range.	83
What is a <u>primate city</u>?	According to Mark Jefferson's "law," a country's largest city that is simultaneously most expressive of the national culture and often the capital city as well; you should be able to name several (clue: try France, Sweden, Austria, and Poland).	83
How do European cities differ from U.S. cities?	A careful re-reading of Lawrence Sommers' quotation underscores the essential differences, centering around age, land availability, investment capital, real-estate and rent control, planning and zoning enforcement, and two different cultural traditions as to living in cities.	83-84
Define <u>central business district</u> (<u>CBD</u>).	A city's heart, its downtown that is marked by high land values, tall buildings, and the clustering of business, commerce, and government.	84
What is the New Towns Movement?	Part of the continentwide urban planning crusade to avoid American-style uncontrolled suburbanization by dispersing people and activities to a number of self-sufficient new satellite communities connected to the nearby city by good public transportation.	84

THE BRITISH ISLES	THE BRITISH ISLES	
What are the political components of this region?	Britain, containing England, Wales, and Scotland, and Ireland, containing the Irish Republic (Eire) and Northern Ireland; all but Eire form the United Kingdom.	85
How do Highland and Lowland Britain differ?	Highland Britain is mainly comprised by Wales and Scotland; Lowland Britain by England except for the Pennines.	85
Where did England's Industrial Revolution originate and flower?	In the Midlands arc surrounding the southern Pennines where coal was plentiful (and in the Northeast too, though to a lesser degree); London was not directly involved.	87
What is Britain's present energy situation?	Whereas coal has declined sharply (and is now imported), oil and gas from beneath the nearby North Sea have now given the U.K. a new self-sufficiency in energy supply for the foreseeable future.	87
Is Britain the world's most highly urbanized society?	Yes--about 90 percent of its population is urbanized which puts it among the leaders; at the same time, agriculture is declining and more food than ever is imported (more than in any other major country).	88
Define <u>conurbation</u>.	General term denoting a large multi-metropolitan complex formed by the coalescence of two or more major urban areas; the Boston-Washington <u>Megalopolis</u> is an outstanding example (see text p. 30).	89
What is London's Green Belt?	Part of the planning movement, a ring of preserved open space surrounding London as shown in Fig. 1-16 (p. 89).	89
To which region does Cornwall belong? Birmingham, England's second city? Newcastle? Liverpool and Manchester? The scenic Lake District?	Respectively, the Southwest, the Midlands, the Northeast, the Northwest (Lancashire), Northern England.	89-90
Which two nearby cities are the industrial and cultural focii of Scotland?	Respectively, Glasgow and Edinburgh.	91

Both Scotland and Wales are potential candidates for <u>devolution</u>--define this concept.	Dévolution, in political geography, is the disintegration of a nation-state due to emergent or resurgent nationalism.	91-92 box
Who are the contestants for power in Ulster (Northern Ireland)?	The Protestant majority who comprise two-thirds of the residents vs. the 35 percent who are Roman Catholic; the British are trying to prevent a civil war.	91-92
WESTERN EUROPE	WESTERN EUROPE	
Which countries comprise this region?	France, West Germany, Belgium, the Netherlands, Luxembourg, Switzerland, and Austria.	94
How long have East and West Germany been partitioned?	Since the end of World War II (1945); the Soviets have blocked all attempts at reunification--see box on similarly fragmented Berlin (text pp. 99-100).	94
What is the <u>Ruhr</u>? Where is its leading port outlet?	Europe's leading industrial complex, based on superb coal supplies and located near the lower Rhine Valley in western West Germany; its port is Rotterdam, located across the Dutch border at the mouth of the Rhine.	94
What is the difference between <u>site</u> and <u>situation</u>?	<u>Site</u> refers to the internal, local physical and other attributes of a place; <u>situation</u> is the regional position of a place or its external locational qualities.	96
How are the economies of Belgium and the Netherlands complementary to each other?	Belgium is a dominantly industrial country producing a large surplus of varied manufactured goods; the Netherlands is a largely agricultural country, producing substantial surpluses of several food commodities.	99-100
What is <u>Randstad</u>?	The Dutch "ring-city" conurbation linking Amsterdam, The Hague, Rotterdam, and Utrecht.	101
Which international organizations are headquartered in Brussels?	The Common Market (European Community) and the North Atlantic Treaty Organization (NATO).	102
What are the four languages spoken in Switzerland? Name another Western European country with more than one official language.	German, French, Italian, and Romansch; Belgium, whose Flemish residents speak a variety of Dutch, and whose Walloon residents speak French.	103

Question	Answer	Page
What city and river are synonymous with Austria?	Vienna and the Danube, which these days is iron gray not blue.	105
NORDIC EUROPE (NORDEN)	NORDIC EUROPE (NORDEN)	
Which countries comprise this region?	Norway, Denmark, Sweden (collectively called Scandinavia)--also Iceland and Finland.	105
Which Scandinavian country is a major dairy-product exporter?	Denmark, whose milder climate and undulating terrain are especially favorable.	105
Why is Sweden so much colder than other European countries at its latitude?	The Norwegian mountains block off milder maritime Atlantic air, allowing Arctic conditions to penetrate southward.	107
How is acid rain formed? How does it affect Norden?	It forms when sulfur dioxide and nitrogen oxides--caused by burning oil, coal, and natural gas--combine with water vapor to produce acidic precipitation; prevailing winds carry these pollutants to Scandinavia, especially southern Norway, where deforestation and fish kills are common.	107-108 box
Which country is Norden's leading wood producer?	Finland.	109
How have the seas favored Norway in recent years?	Huge deposits of oil and natural gas in the Norwegian sector of the North Sea have begun yielding both energy and heigtened export income.	110-box
MEDITERRANEAN EUROPE	MEDITERRANEAN EUROPE	
Which countries comprise this region?	Greece, Italy, Spain, and Portugal.	111
Which countries are located on the Iberian Peninsula?	Spain and Portugal.	111
Which country is the region's largest hydroelectric power producer?	Italy, one of the world's leading countries (Norway is another) for this form of energy production.	112
Which vital subregion of Italy contains nearly half its population and today functions as its economic core?	The Po River basin of the north, which contains the cities of Milan, Turin, Genoa, and Venice, and also belongs to Europe's core area (Fig. 1-11, p. 75).	112, 116-117

What generalization can be made about the population distributions of this region's countries?	They are decidedly peripheral with large concentrations in productive coastal and riverine lowlands.	113-114
How do the economies of this region compare to those to the north?	They lag considerably, emphasizing less-productive agriculture.	114
Where is Spain's major manufacturing zone located?	In Catalonia in the northeast, focused on Barcelona.	115
Why is Italy Mediterranean Europe's leading state?	Because of its contribution to Western culture, more productive agriculture and manufacturing, and higher living standards.	115-116

EASTERN EUROPE	EASTERN EUROPE	
Which countries comprise this region?	Poland, Czechoslovakia, Hungary, Yugoslavia, Romania, Bulgaria, and Albania.	117
What is meant by the term hegemony?	The political dominance of a region or country by another country; the Soviet Union's grip on Eastern Europe is a classic example.	117
What is irredentism?	A policy of cultural extension and potential political expansion aimed at a national group living in a neighboring country.	119
What is meant by the term balkanization? Why is this region so often called a shatter belt?	The fragmentation of a region into smaller, often hostile, political units; because of the numerous cultures that have collided here, the resultant shattering of various cultural and political units has produced an especially fragmented ethnic map--held together (for now) by Soviet hegemony.	120-box, 121
Where is Poland's leading industrial area located?	Silesia is located in southern/southwestern Poland, anchored by the cities of Krakow and Wroclaw.	120
What is the significance of the Bohemian Basin?	Czechoslovakia's core area; the zone around Prague is one of the region's leading manufacturers and a cosmopolitan center with a rich history.	121

Question	Answer	Page
Why is the Danube a failure at unifying Eastern Europe?	Apart from the region's political kaleidoscope of rivalries, the river is almost a constant dividing line and did not attract the core areas of many countries it flows through to locate in its valley.	122
Which country is Eastern Europe's leading federal state?	Yugoslavia, one of the world's most ethnically diverse countries (see Fig. 1-29 on p. 118).	123
EUROPEAN UNIFICATION	EUROPEAN UNIFICATION	
What is meant by the term <u>supranationalism</u>?	International cooperation involving the voluntary participation of 3 or more countries in an economic, political, or cultural association.	126
What are Benelux and OEEC?	Europe's first efforts at unification: Benelux (1944) involved the economic association of the Low Countries and Luxembourg, OEEC (1948) was the Organization for European Economic Cooperation set up by the early postwar U.S.-aided Marshall Plan.	126-127
Name the countries of the EC (the Common Market), the European Free Trade Association, and COMECON.	<u>EC</u>: France, United Kingdom, West Germany, Italy, Greece, Denmark, Ireland, Spain, Portugal, and the 3 Benelux countries; <u>EFTA</u>: Sweden, Norway, Switzerland, Austria, Iceland (and Finland as associate member); <u>COMECON</u>: the U.S.S.R. and all of its Eastern European satellites except Albania (Yugoslavia is only a partial member).	127-128

MAP EXERCISES

MAP COMPARISON

1. Compare the maps of London (Fig. 1-16, p. 89) and Paris (Fig. 1-19, p. 98); what similarities can be observed in their spatial structuring and land-use patterns?

2. Compare the map of Europe's relative world location (Fig. 1-5, p. 63) with the world political map (Fig. I-17, pp. 34-35); discuss the differences between the relative and absolute location of Europe, and how different map projections can convey very different perspectives.

3. Compare Europe's population distribution (Fig. 1-2, p. 61) with the world map of the same phenomenon (Fig. I-14, pp. 26-27), and list the features of the European map that only become clear at the larger scale; also, compare Fig. 1-2 to the world cartogram (Fig. I-15, pp. 28-29) and make additional comparisons about the relative internal variation of Europe's population geography.

4. How does Europe's population distribution correspond to the realm's physical landscape? Compare Figs. 1-2 (p. 61) and 1-6 (p. 65) in order to make your observations, paying particular attention to the location of cities and dense population corridors.

5. The boundary of Europe's core area is very carefully drawn in Fig. 1-11 (p. 75). Briefly trace that boundary through each country it lies in and, by referring to other maps and appropriate text, justify why the line follows the route it does.

6. By referring to other maps and pertinent text, explain why the acid rain distribution (Fig. 1-25 on p. 108) assumes the pattern, density, and dispersion it does within the portion of Europe shown on this map.

7. The languages of Europe perform a variable function in the realm's political units, sometimes binding people together and sometimes dividing them further. By reviewing Fig. 1-13 (p. 81) and appropriate text, comment on the status of language as a political force in each of the following countries: United Kingdom, Belgium, Switzerland, Poland, Yugoslavia, Finland, and Romania.

MAP CONSTRUCTION

(Use outline maps at the end of this chapter)

1. In order to familiarize yourself with Europe's physical geography, place the following on the first of the outline maps:

   a. Rivers: Thames, Seine, Loire, Rhone, Garonne, Meuse (Maas), Rhine, Elbe, Danube, Po, Ebro, Tagus, Guadalquivir, Vistula (Wisla), Oder (Odra), Tisza, Sava

   b. Water bodies: Mediterranean Sea, Baltic Sea, North Sea, Irish Sea, English Channel, Straits of Gibraltar, Tyrrhenian Sea, Adriatic Sea, Ionian Sea, Zuider Zee

c. <u>Land bodies</u>: Ireland, Sicily, Corsica, Sardinia, Balearic Islands, Shetland Islands, Jutland Peninsula, Peloponnesus

d. <u>Mountains</u>: Alps, Pyrenees, Appennines, Pennines, Scottish Highlands, Dolomites, Tatras, Ore Mountains (Erzgebirge), Sudeten Mountains, Jura, Cantabrians, Carpathians, Balkans, Transylvanian Alps

e. <u>Other landforms</u>: Meseta, Po Plain, Massif Central, North European Lowland, Bohemian Basin

2. On the second map, political-cultural geographic information should be entered as follows:

   a. Label each country that is included in the Appendix A table listing under Europe (text p. 611)

   b. For each of those countries locate and label the capital city with the symbol *

   c. Reproduce the language map using light pencil coloring (Fig. 1-13, p. 81).

3. On the third outline map, economic-urban information should be entered as follows:

   a. <u>Europe's regions</u>: Using a thick line, draw the boundaries of the British Isles, Western Europe, Nordic Europe, Mediterranean Europe, and Eastern Europe.

   b. <u>Supranational affiliations</u>: Color the countries of the Common Market (EC) green, countries belonging to the European Free Trade Association yellow, and countries that are members of COMECON red (include associate members too)

   c. <u>Cities</u> (locate and label with the symbol ·): London, Birmingham, Manchester, Liverpool, Newcastle, York, Leeds, Edinburgh, Glasgow, Dublin, Belfast, Paris, Marseille, Lyon, Bordeaux, Toulouse, Strasbourg, Antwerp, Brussels, Luxembourg, Rotterdam, Amsterdam, Hamburg, Cologne, Essen, Dusseldorf, Frankfurt, Berlin, Stuttgart, Munich, Zurich, Geneva, Vienna, Graz, Copenhagen, Oslo, Bergen, Goteborg, Stockholm, Helsinki, Milan, Turin, Genoa, Venice, Florence, Rome, Naples, Palermo, Madrid, Lisbon, Seville, Barcelona, Athens, Istanbul, Warsaw, Krakow, Katowice, Prague, Budapest, Bucharest, Sofia, Trieste, and Belgrade

   d. <u>Economic regions</u> (identify with circled letter):

      A - the Midlands
      B - East Anglia
      C - "Waist" of Scotland
      D - the Ruhr
      E - Paris Basin
      F - Silesia
      G - Randstad-Holland
      H - Catalonia
      I - Lombardy
      J - Bohemia

PRACTICE EXAMINATION

SHORT-ANSWER QUESTIONS

Multiple-Choice

1. Which of the following mountain ranges is regarded by a number of geographers as Europe's eastern boundary:

    a) Alps
    b) Appennines
    c) Pyrenees
    d) Urals
    e) Carpathians

2. Which of the following cities is located in Italy's core area:

    a) Milan
    b) Rome
    c) Barcelona
    d) Naples
    e) Catalonia

3. Which of the following capital cities is not a primate city:

    a) Paris
    b) Lisbon
    c) Athens
    d) Bonn
    e) Vienna

True-False

1. The Republic of Ireland (Eire) is not a part of the state called the United Kingdom.

2. Although it pursued the acquisition of territory and precious metals, mercantilism was not concerned with actively spreading Christianity throughout the New World.

3. The spatial interaction principle of transferability refers to the ability to move a good at a bearable cost.

Fill Ins

1. The Isolated State model of commercial agricultural spatial organization was devised by the economist ______________.

2. The chief landform feature of the Iberian Peninsula is a tableland called the ______________ Plateau.

3. The frequency of occurrence of a phenomenon within a given area is known as its ______________.

Matching Question on European Countries

_____	1.	Seat of the World Court	A.	Austria
_____	2.	Midi grape-growing	B.	Belgium
_____	3.	Welsh devolution threat	C.	Bulgaria
_____	4.	Home of Pilsner beer	D.	Czechoslovakia
_____	5.	Ekofisk and Statfjord	E.	Denmark
_____	6.	Appennine Mountains	F.	France
_____	7.	Borders Black Sea	G.	East Germany
_____	8.	International conference headquarters	H.	Hungary
_____	9.	Serbo-Croatian cooperation	I.	Italy
_____	10.	Jutland Peninsula	J.	The Netherlands
_____	11.	Southwestern Iberia	K.	Norway
_____	12.	Forced to join Nazi Germany in 1938	L.	Portugal
_____	13.	Home of the Magyars	M.	Switzerland
_____	14.	Headquarters of NATO	N.	United Kingdom
_____	15.	West Berlin	O.	Yugoslavia

ESSAY QUESTIONS

1. One of the major turning points in European (and world) history was the achievement of the Industrial Revolution. Discuss the sequence of events that marked this economic transformation, highlight its geographical expressions, and review the stages of the Demograhic Transition Model and its impact on population trends in Britain over the past two centuries.

2. One of the ten major geographic qualities of Europe is its strong regional differentiation, which has given rise to a high degree of areal functional specialization affording multiple exchange opportunities. For three of the realm's five major regions, provide an example of such spatial complementarity and discuss the interaction that it has produced during this century.

3. Review the principles that underlie the Von Thunen model, and discuss the various applications of this generalization to Europe in Von Thunen's day as well as the 20th century.

4. The partitioning of Germany has been one of Europe's major politico-geographical events in this century, yet both East and West Germany have experienced economic growth since 1945. Discuss the resource base and economic-geographical infrastructure of each country, and show how the West Germans have made the most of their opportunities to become one of the world's leading economic powers during the past three decades.

5. Eastern Europe, the theater of cultural collisions and conflicts for centuries, is a classic example of a shatter belt. Discuss the concept of the "shatter belt" and the balkanization it has produced here, and give several examples of persistent cultural fragmentation in the region today despite the straitjacket of Soviet hegemony.

TERM PAPER POINTERS

There is a great deal of good, accessible literature on the European realm. A good place to start your research, once you and your instructor have agreed on a topic to pursue, is an up-to-date regional geography of Europe. Several excellent texts exist, and many are of very recent vintage. Among them are (see reference listing on text pp. 130-131): Diem, Hoffman [1983], Ilbery, Jordan, Mellor & Smith, and Monkhouse--a particularly fine older book is the one by Gottmann. Several such books also focus on individual regions (Beckinsale & Beckinsale, Clout, [1986], Clout et al., Demko, Hudson & Lewis, Knox, and Russ) and on specific countries (Boal & Douglas, Evans, Doornkamp, Rugg Mellor [1978], Orme, Pinchemel et al., Pringle, Short & Kirby, and Wild). The best approach is to read up on the topic or region in the appropriate work, being careful to spot additional references to your subject; maps are also important, and should be found in these works as well as in specialized atlases on Europe (see, for instance, Clayton & Kormoss). You may also find more about your subject in topical geographies of the realm and its components--some examples are Chapman, Embleton, Foster, Hall & Ogden, Houston [1963], Kosinski, Manners, Nystrom & Hoffman, Parker and Pounds. And by all means, do not overlook current-events sources; the article by Markham is an example of the latest reporting--the New York Times Index is an especially good place to begin, and periodicals such as Focus (see Bibliographical Footnote on p. 131) and Geographical Magazine are other sources to keep in mind. Finally, be aware that if urbanization in Europe is related to your topic, several useful recent references are listed in conjunction with Chapter One's subtheme on population geography: Burtenshaw et al., Glebe & O'Loughlin, Hohenberg & Lees, Sommers, and White. You should also review the term-paper section in the previous chapter, which introduces you to the professional journals of regional geography.

The population geography Systematic Essay also provides numerous opportunities to develop research papers, either on this topic itself or applied directly to Europe. Again, a good systematic geography is a fine place to start, giving you background on your topic and steering you toward other relevant literature. The best surveys are those by Newman & Matzke and Jones; other useful sources are Clarke, Demko et al., Hall & Ogden, Kosinski, Lee, Petersen & Petersen, and Zelinsky. This subject is also dealt with in a number of other places in the text, especially in Chapter 8 on pp. 498-502. The topic also is repeatedly covered in the news media--the 1984 U.N. world conference in Mexico City received much attention. An excellent report that summarizes global population trends in the mid-1980s is World Development Report 1987, published by Oxford University Press for the World Bank (1987). Population data are also very important for research on this topic; by now you should be familiar with the data displayed in Appendix A (text pp. 611-614), and there is much more information available in the publication from which the table was drawn (see reference in the book preface) that can be found in the Government Documents section of your library.

SPECIAL EXERCISE

Model Box 2 (pp. 76-78) treats the Von Thunen model which can be a helpful approach to understanding the spatial organization of agricultural activity. This special exercise asks you to find one of two articles in the library, read it, and make certain observations. It is advised that you prepare yourself by reviewing a more extended elaboration of Von Thunen's principles; your instructor can recommend several sources (two appropriate places are references cited on pp. 130-131--either Chisholm, Chapters 1 and 2, or Wheeler & Muller, Chapter 13). The two articles which you may choose from are:

1. Harvey, David. "Locational Change in the Kentish Hop Industry and the Analysis of Land Use Patterns," Institute of British Geographers, Transactions, 33(1963), 123-144. This study is also reprinted in Smith, Robert H.T., Taaffe, Edward J., and King, Leslie J., eds., Readings in Economic Geography: The Location of Economic Activity (Chicago: Rand McNally, 1968), pp. 79-93.

2. Griffin, Ernst. "Testing the Von Thunen Theory in Uruguay," Geographical Review, 63 (October 1973), 500-516.

If you choose the Harvey article, comment upon the following:

a. are physical-geographical forces sufficient to explain the distribution of hop cultivation in Kent?

b. list five agglomerative advantages ("external economies of scale") that encouraged the concentration of hops in central Kent

c. how did landlords help to assure that the region specialized in hops, the most profitable crop in accordance with Thunian principles?

If you select the Griffin article, comment upon the following:

a. why is the country of Uruguay a particularly good empirical "laboratory" in which to test the Von Thunen theory?

b. compare and contrast the theoretical Von Thunen map for Uruguay (left-hand map on p. 510 of journal) with the one for Europe (textbook p. 77)

c. summarize the similarities between model and reality for Uruguay--and do you agree with the author's conclusions?

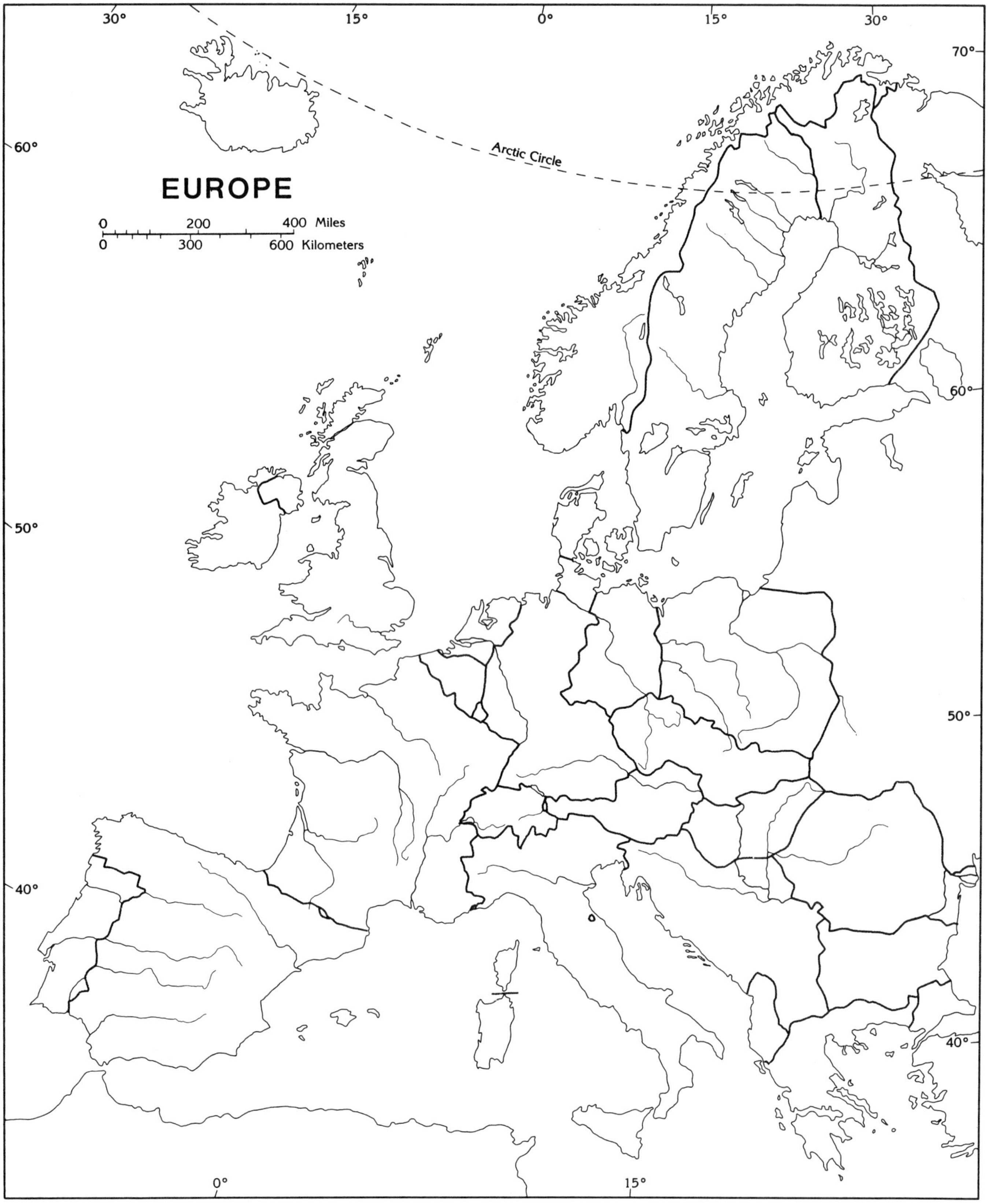
EUROPE
0 200 400 Miles
0 300 600 Kilometers
Arctic Circle
30°
15°
0°
15°
30°
70°
60°
50°
40°
60°
50°
40°
0°
15°

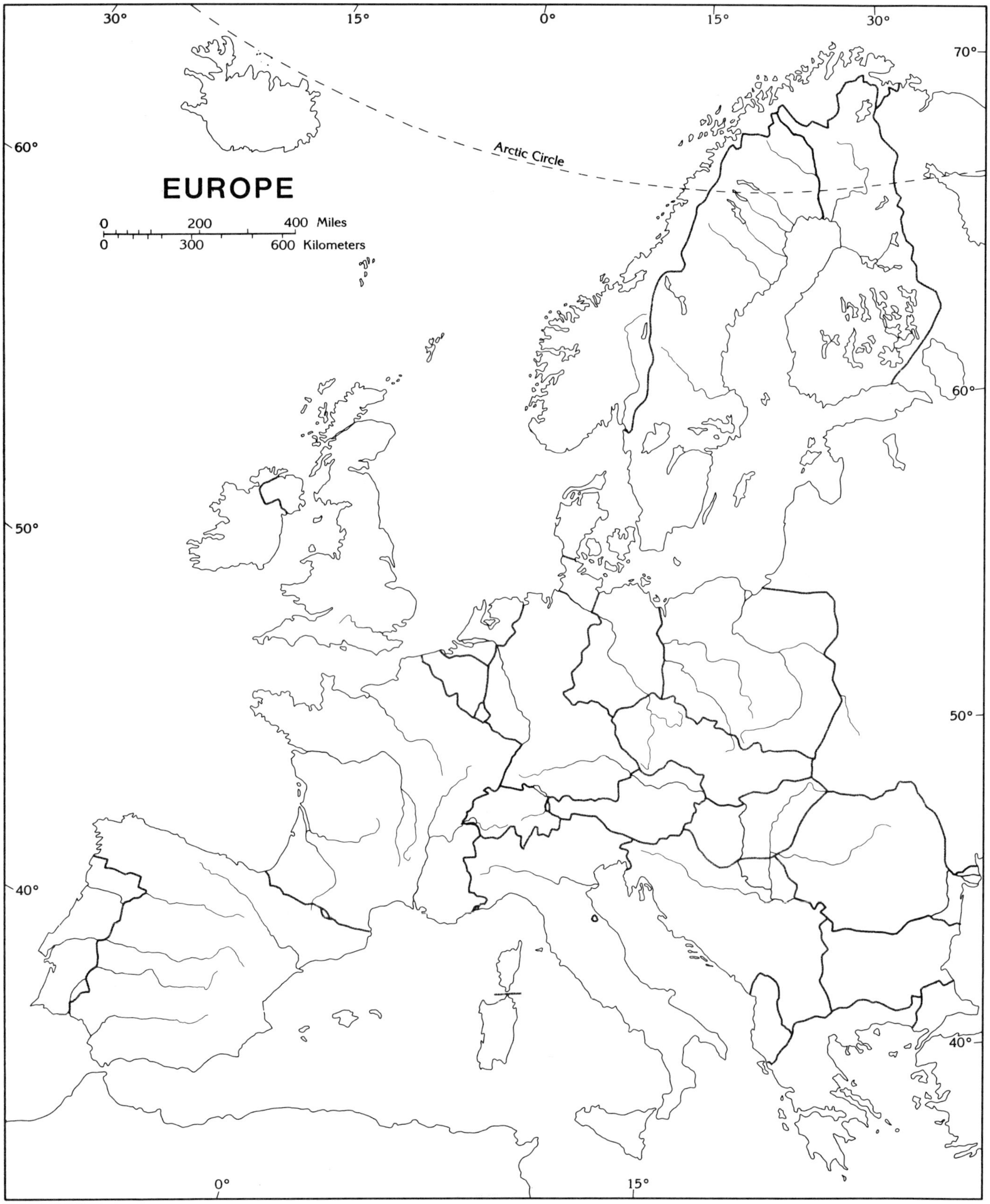

EUROPE
0 200 400 Miles
0 300 600 Kilometers
Arctic Circle
30°
15°
0°
15°
30°
70°
60°
50°
40°
60°
50°
40°
0°
15°

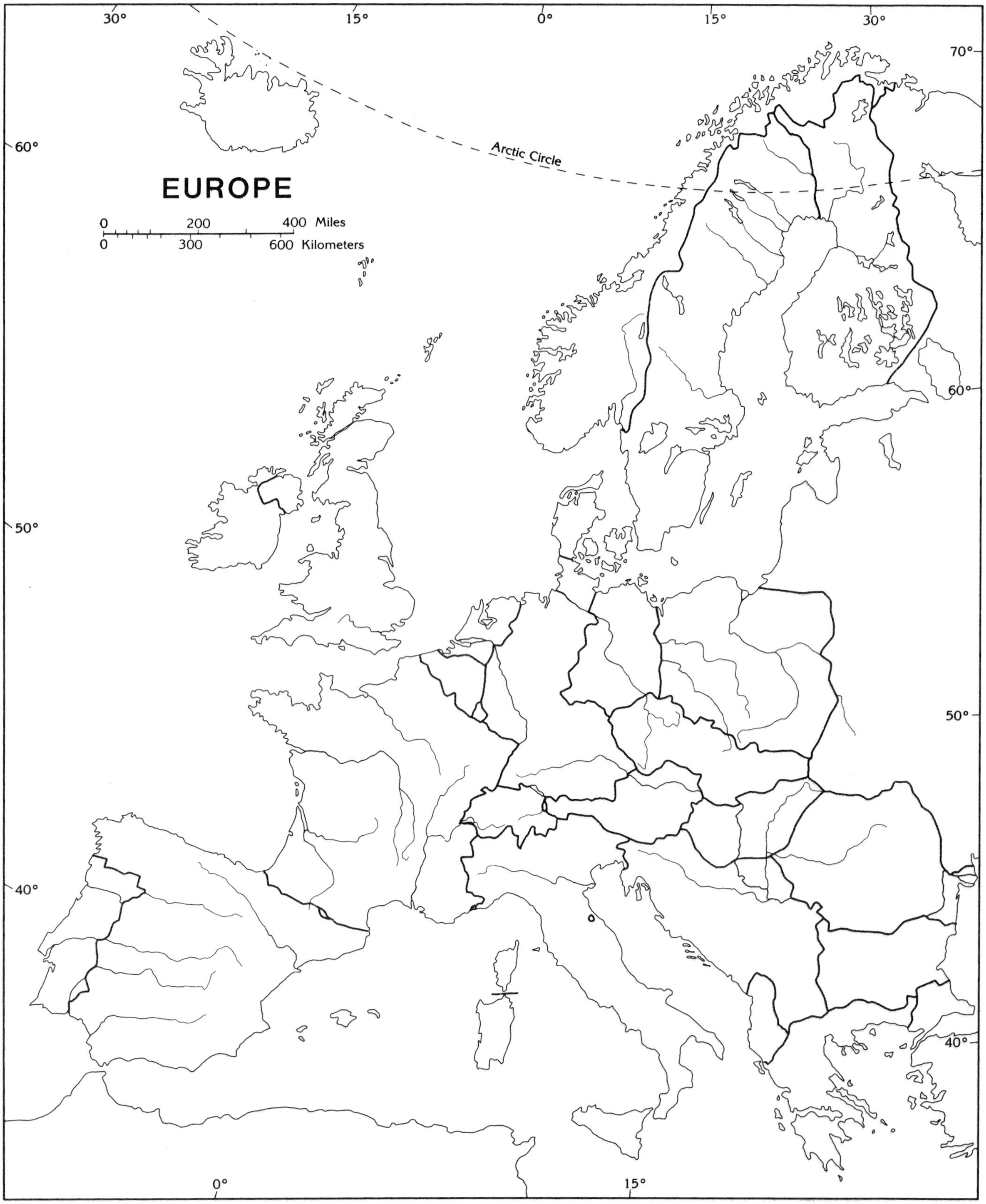

EUROPE
0 200 400 Miles
0 300 600 Kilometers
Arctic Circle
30°
15°
0°
15°
30°
70°
60°
50°
40°
60°
50°
40°
0°
15°

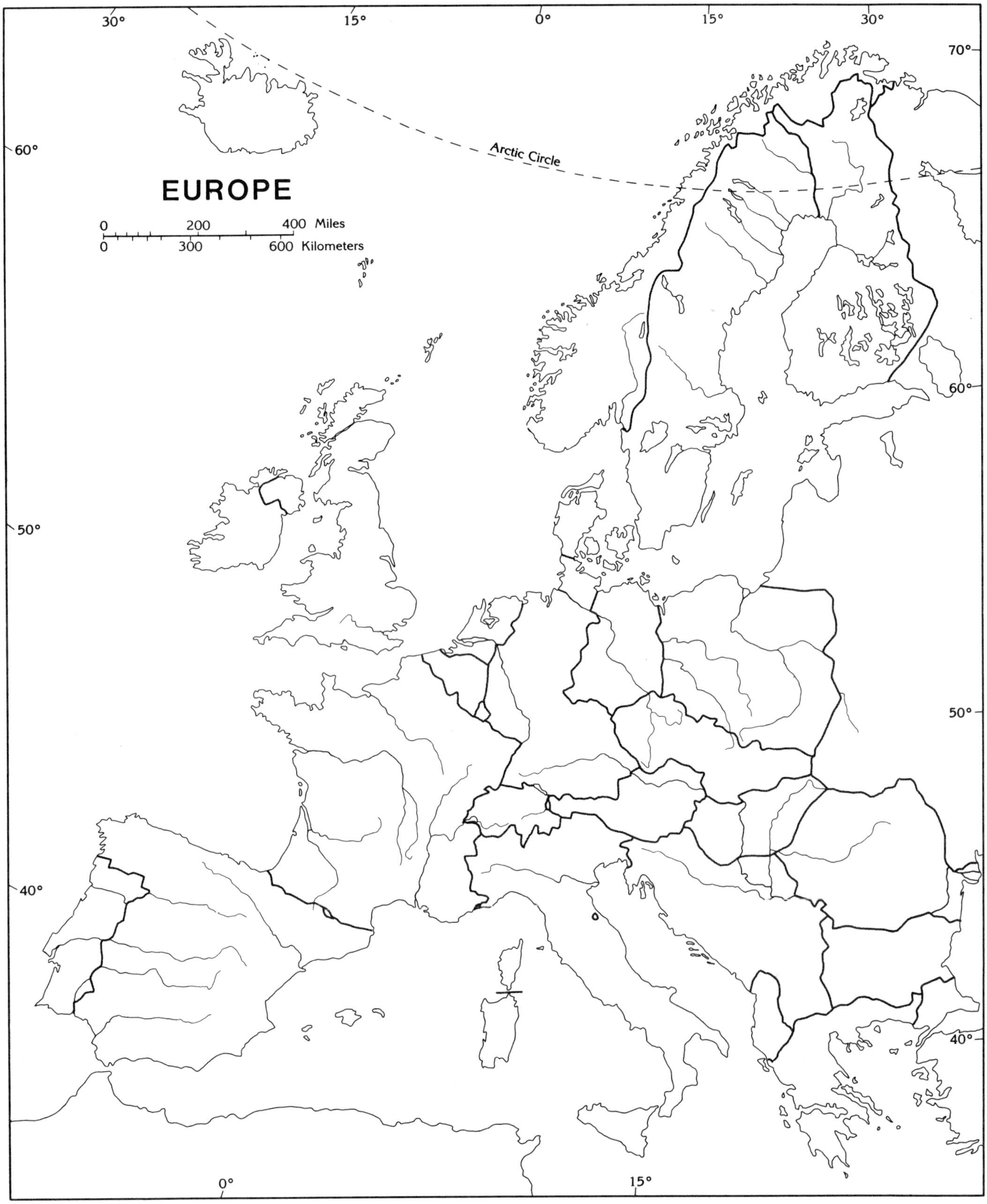
EUROPE
0 200 400 Miles
0 300 600 Kilometers
Arctic Circle
30°
15°
0°
15°
30°
70°
60°
50°
40°
60°
50°
40°
0°
15°

VIGNETTE: AUSTRALIA AND NEW ZEALAND—EUROPEAN OUTPOST

OBJECTIVES OF THIS VIGNETTE

This appendage to Chapter 1, the first of five such Vignettes in the book, treats the Australasian realm which consists of Australia and New Zealand. Our brief survey covers population movements, economic and urban geography, farming, mineral resources, manufacturing, immigration policy, federalism, and close-up view of New Zealand's geographical characteristics.

Having learned the regional geography of this corner of the world, you should be able to:

1. Explain why Australasia is an offshoot of the European realm and therefore unlike its surrounding realms.

2. Understand the process of migration, its various subtypes, and Ravenstein's generalizations.

3. Understand the course of Australia's evolution, with special emphasis on its economic geography.

4. Explain why Australia's agricultural specializations developed, and why certain farming activities located where they did.

5. Identify Australia's mineral resources and their distributions, and their relationship to the country's manufacturing base.

6. Discuss Australia's population policies and how they helped to shape its society.

7. Understand Australia's complex political geography, why federalism has been so successful there, and how the location of its capital city was decided upon.

8. Understand the highlights of New Zealand's geography as profiled on text pp. 145-147.

9. Locate the leading physical, cultural, and economic-spatial features of this realm on an outline map.

GLOSSARY

Migration (134-135)

A change in residence intended to be permanent. Population redistribution of this kind often assumes flows or streams, composed of migrants making similar relocation decisions; in the U.S. today, for instance, such flows extend from the north to the Sunbelt, from cities to suburbs, and, for the elderly, toward retirement regions (again in the Sunbelt). Ernst Ravenstein's generalizations have led to the establishment of a large body of modern theory in which such forces as environmental perception, information flow, and distance play important roles in shaping individual decision making.

Unitary state (143)

Deriving from the Latin word for "one." A unified and centralized state, usually with a long tradition of being ruled by a single authority, in which power is exercised equally throughout the country.

Federal state (143)

Deriving from the Latin word for "association." A political framework whereby a central government represents the various entities (Australia's 6 states and 2 territories in this case) within a nation-state where they have common concerns--defense, foreign affairs, and the like--yet permits these entities to retain their own identities and to have their own laws, policies, and customs in certain spheres.

Compromise capital (143-144)

A capital city that represents a compromise solution when several cities vie for the honor of becoming the administrative headquarters. Usually involves formation of a new federal district which is separate from pre-existing political units. Canberra is a classic example--resolving the late 19th century struggle between Sydney and Melbourne to become Australia's capital; another compromise capital is Washington, D.C. in the United States.

SELF-TESTING QUESTIONS

Cover this side of the page with a sheet of paper. Uncover each line after you have attempted to answer the question in the left column. If necessary, refer to the textbook page(s) listed at the right.

Question	Answer	Page
AUSTRALASIA'S CHARACTERISTICS	AUSTRALASIA'S CHARACTERISTICS	
Why is Australia-New Zealand a discrete, European-related realm?	Its population and culture are decidedly European, and its livelihood sharply contrast to those of surrounding realms; Australasia is also large enough to rank with any continental land mass.	132
How did Australia's indigenous peoples arrive on the island continent?	By way of a land bridge that connected to Southeast Asia by way of New Guinea when Pleistocene sea levels were lower.	133
MIGRATION	MIGRATION	
Name three factors associated with the intensity of a migration flow.	(1) Perceived difference between one's origin and destination; (2) information sent back by earlier migrants; (3) distance.	134
How does the concept of intervening opportunity affect migration behavior?	People's perceptions of faraway destinations are less positive when there are closer opportunities.	134
What push and pull factors attracted people to migrate from England to Australia?	Pull: a new frontier, land availability, and the chance to make a fortune; push: crowded conditions, and a less favorable climate.	135
Who compromised most of Australia's earliest European settlers?	Convicts: prisoners who were forcibly deported from Britain to this remote penal colony.	135
ECONOMIC GEOGRAPHY	ECONOMIC GEOGRAPHY	
What is meant by the term "outback"?	The dry, sparsely populated Australian interior.	135

What percentage of today's Australians reside in urban areas on or near the coast?	About 86 percent.	136
How well was Australia's transport network integrated?	Fierce competition among its states slowed national integration; because of railroad gauge differences, it was 1970 before trains could travel from Sydney to Perth on the same width of track.	137
How much of Australia's population resides in the metropolitan areas of Sydney and Melbourne?	About 7,000,000 of the country's 16,000,000 inhabitants.	137
What is Australia's leading commercial crop?	Wheat, which earns about half of the annual income derived from sheepraising.	139
Where are Australia's main dairying zones located?	Close to the urban markets of the moister coastal areas.	139
What kind of a spatial pattern is exhibited by Australia's varied mineral deposits?	Scattered across the country, but in high enough concentrations to make mining profitable.	139-140
What is Australia's most important energy resource?	Coal, which is used to produce about 90 percent of the country's electricity.	140
Which minerals are produced at Broken Hill?	This is one of the world's leading areas for lead-and zinc-mining.	141
What is the present condition of Australian manufacturing?	Serious and declining; high production costs and long distances to foreign markets inhibit production, and force the importation of many goods that could be produced domestically.	141
POPULATION POLICIES	POPULATION POLICIES	
Which ethnic groups have entered Australia under its policy of selective immigration, and which have not?	Favored have been the British, Germans, Dutch, Greeks, and Turks; nonwhites and Asians have been severely restricted.	141-142
How has Australia reacted to the problem of Vietnamese and Kampuchean "boat-people" refugees?	Surprisingly, more than 50,000 have been admitted.	142
How were Australian immigration procedures modified in 1979?	Quotas were liberalized as were admission requirements, but the inflow is still tightly controlled; immigrant unemployment, however, has risen sharply.	142

POLITICAL GEOGRAPHY	POLITICAL GEOGRAPHY	
Why is Australia a <u>federal</u> <u>state</u>?	Because it was formed as a union composed of several separated colonies who wanted to preserve their identities and customs.	143
How was a capital city chosen in such a competitive environment?	By compromise: Canberra, about halfway between Sydney and Melbourne, was created within a new federal district in 1908.	143
NEW ZEALAND	NEW ZEALAND	
Who are the pre-European people of New Zealand, Polynesians who are still present in large numbers?	The Maoris.	145
What is the largest agricultural region of the South Island?	The Canterbury Plain.	145
What do Australians and New Zealanders share?	A joint British heritage, a substantially pastoral agricultural sector, small local markets, rugged interiors, the problem of vast distances to world markets, a high degree of urbanization, peripheral development, and a lagging manufacturing base.	145-146

MAP EXERCISES

MAP COMPARISON

1. Compare the map of modern human migrations (Fig. A-2, p. 134) with the current world population distribution (Fig. I-14, pp. 26-27). How do the planet's four largest population concentrations relate to migration flows: are they leading origins and/or destinations for migrants, and how does European emigration compare to population outflows from the other three largest clusters?

2. Compare the map of Australian minerals and agricultural areas (Fig. A-6, p. 140) with the country's population distribution (Fig. I-14, p. 27). What spatial associations are apparent, and for which activities is remoteness a serious problem? For the latter, has the distance problem been ameliorated by the extension of the national transport network (shown in Fig. A-5, p. 138)?

MAP CONSTRUCTION

(Use outline maps at the end of this Vignette)

1. To familiarize yourself with Australasia's physical geography, place the following on the first of the outline maps:

   a. Rivers: Murray, Darling, Murrumbidgee, Waikato

   b. Water bodies: Bass Strait, Great Australian Bight, Gulf of Carpentaria, Coral Sea, Tasman Sea, Cook Strait

   c. Land areas: Great Dividing Range, Murray Basin, Nullarbor Plain, Great Artesian Basin, Great Sandy Desert, Great Victoria Desert, Arnhem Plateau, Gibson Desert, Tasmania, Southern Alps, Canterbury Plain, North Island, South Island.

2. On the second outline map, enter the following political information:

   a. identify and label the two national capitals with the symbol *

   b. label all of the states and territories of Australia, and label each of their capitals with the symbol +.

3. On the third outline map, enter the following urban-economic information:

   a. Cities (locate and label with the symbol •): Sydney, Melbourne, Adelaide, Brisbane, Canberra, Hobart, Newcastle, Rockhampton, Alice Springs, Darwin, Perth, Kalgoorlie, Wellington, Auckland, Christchurch, Dunedin

   b. Economic centers (identify with circled letter):

      A - Broken Hill
      B - Kambalda
      C - Mount Isa
      D - Canterbury Plain

PRACTICE EXAMINATION

SHORT-ANSWER QUESTIONS

Multiple-Choice

1. Australia's governmental structure can best be classified as a:

   a) unitary state
   b) federal state
   c) Western European satellite
   d) monarchy
   e) dictatorship

2. Which of the following is located on New Zealand's North Island:

a) Auckland
b) Canterbury Plain
c) Adelaide
d) Southern Alps
e) Tasmania

True-False

1. New Zealand's interior is as sparsely populated as Australia's.

2. Wheat is Australia's largest commercial crop.

Fill Ins

1. The capital of New Zealand is ______________.

2. Australia's original function in the British empire was to serve as a ______________.

Matching Question on Australasia

_____	1.	Outback lead-mining center	A.	Maori
_____	2.	Early British explorer	B.	Broken Hill
_____	3.	Devastated by 1974 cyclone	C.	Perth
_____	4.	Outback gold-mining center	D.	New South Wales
_____	5.	Architect of Canberra plan	E.	James Cook
_____	6.	Large Greek population	F.	Walter Burley Griffin
_____	7.	May be largest Polynesian city	G.	Darwin
_____	8.	Integrating into New Zealand society	H.	Kalgoorlie
_____	9.	Capital of Western Australia	I.	Melbourne
_____	10.	Sydney harbor	J.	Auckland

ESSAY QUESTIONS

1. One of the major turning points in European (and world) history was the achievement of the Industrial Revolution. Discuss the sequence of events that marked this economic transformation, highlight its geographical expressions, and review the stages of the Demograhic Transition Model and its impact on population trends in Britain over the past two centuries.

2. One of the ten major geographic qualities of Europe is its strong regional differentiation, which has given rise to a high degree of areal functional specialization affording multiple exchange opportunities. For three of the realm's five major regions, provide an example of such spatial complementarity and discuss the interaction that it has produced during this century.

3. Review the principles that underlie the Von Thunen model, and discuss the various applications of this generalization to Europe in Von Thunen's day as well as the 20th century.

4. The partitioning of Germany has been one of Europe's major politico-geographical events in this century, yet both East and West Germany have experienced economic growth since 1945. Discuss the resource base and economic-geographical infrastructure of each country, and show how the West Germans have made the most of their opportunities to become one of the world's leading economic powers during the past three decades.

5. Eastern Europe, the theater of cultural collisions and conflicts for centuries, is a classic example of a shatter belt. Discuss the concept of the "shatter belt" and the balkanization it has produced here, and give several examples of persistent cultural fragmentation in the region today despite the straitjacket of Soviet hegemony.

TERM PAPER POINTERS

The literature cited on p. 148 is a good introduction to the geography of this realm and to spatial themes that have received attention from geographers. The discipline of geography is well represented in both Australia and New Zealand, and both countries publish journals that many larger libraries subscribe to. One topic that might make an interesting term paper is a comparison of Australian and U.S. geographical problems--several connections are mentioned in the text and they should be easy to follow up on in a research project.

There is no Special Exercise for this Vignette

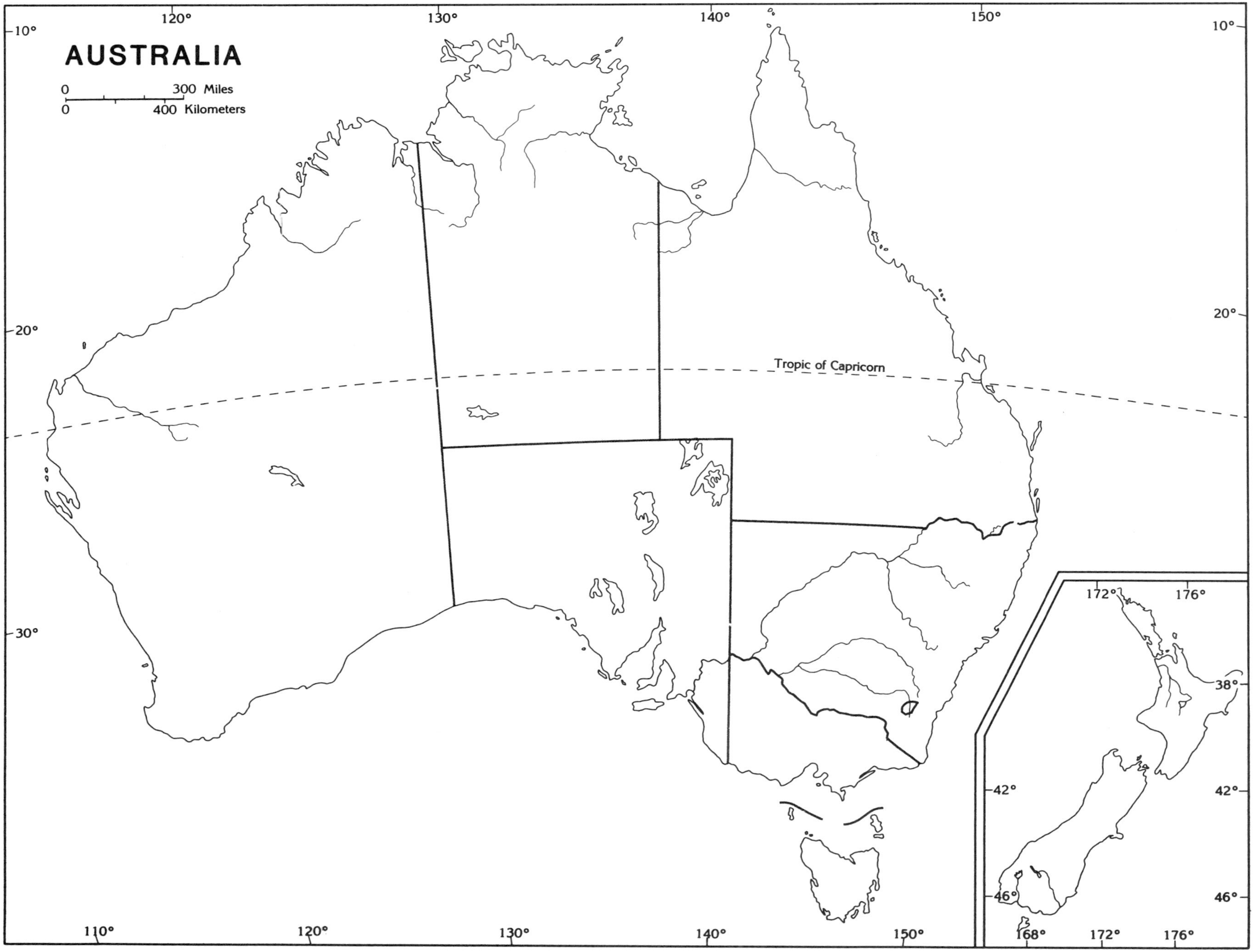
AUSTRALIA
0 300 Miles
0 400 Kilometers
Tropic of Capricorn
10°
20°
30°
110°
120°
130°
140°
150°
172°
176°
168°
38°
42°
46°

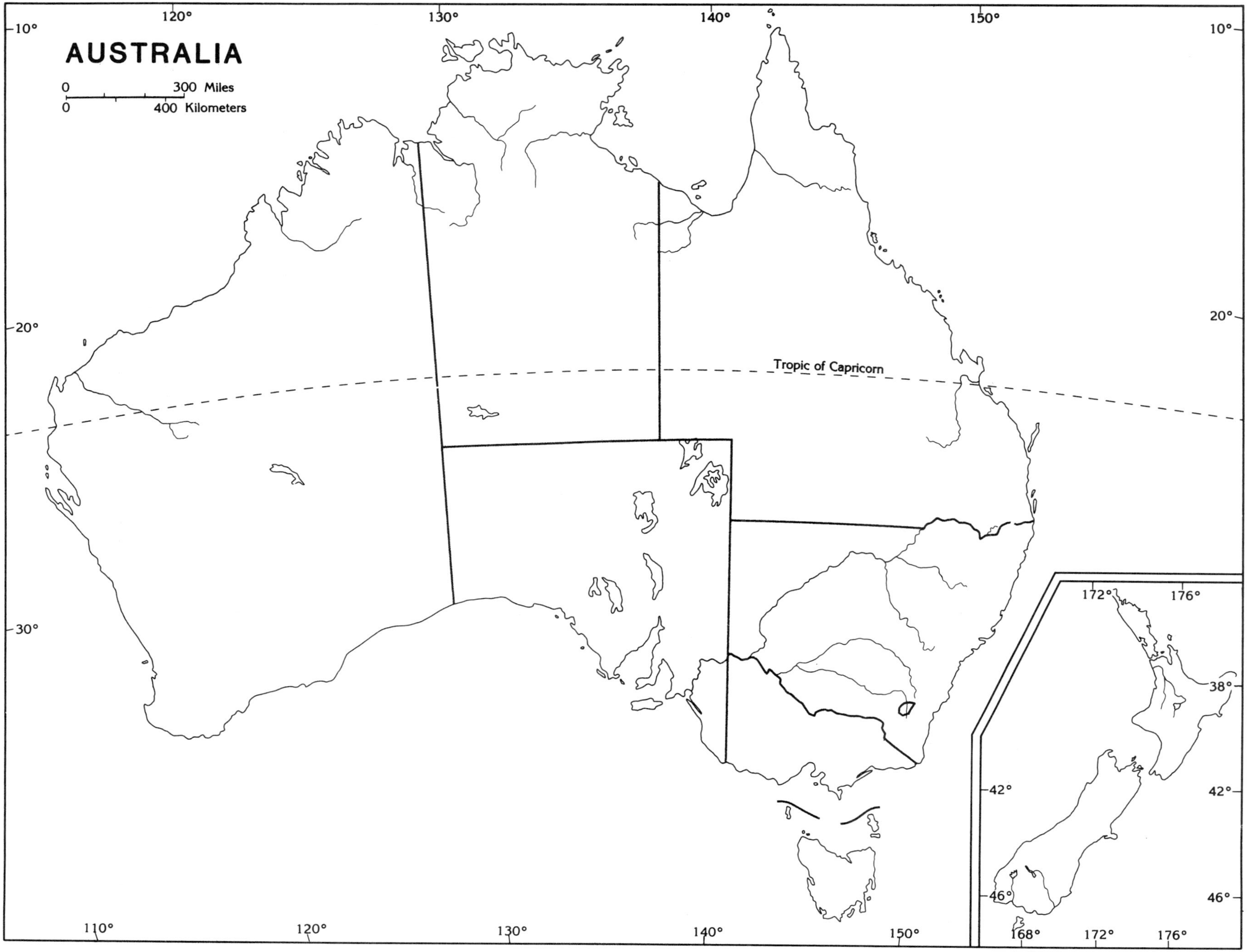

AUSTRALIA
0 300 Miles
0 400 Kilometers
Tropic of Capricorn
10°
20°
30°
110°
120°
130°
140°
150°
172°
176°
168°
38°
42°
46°

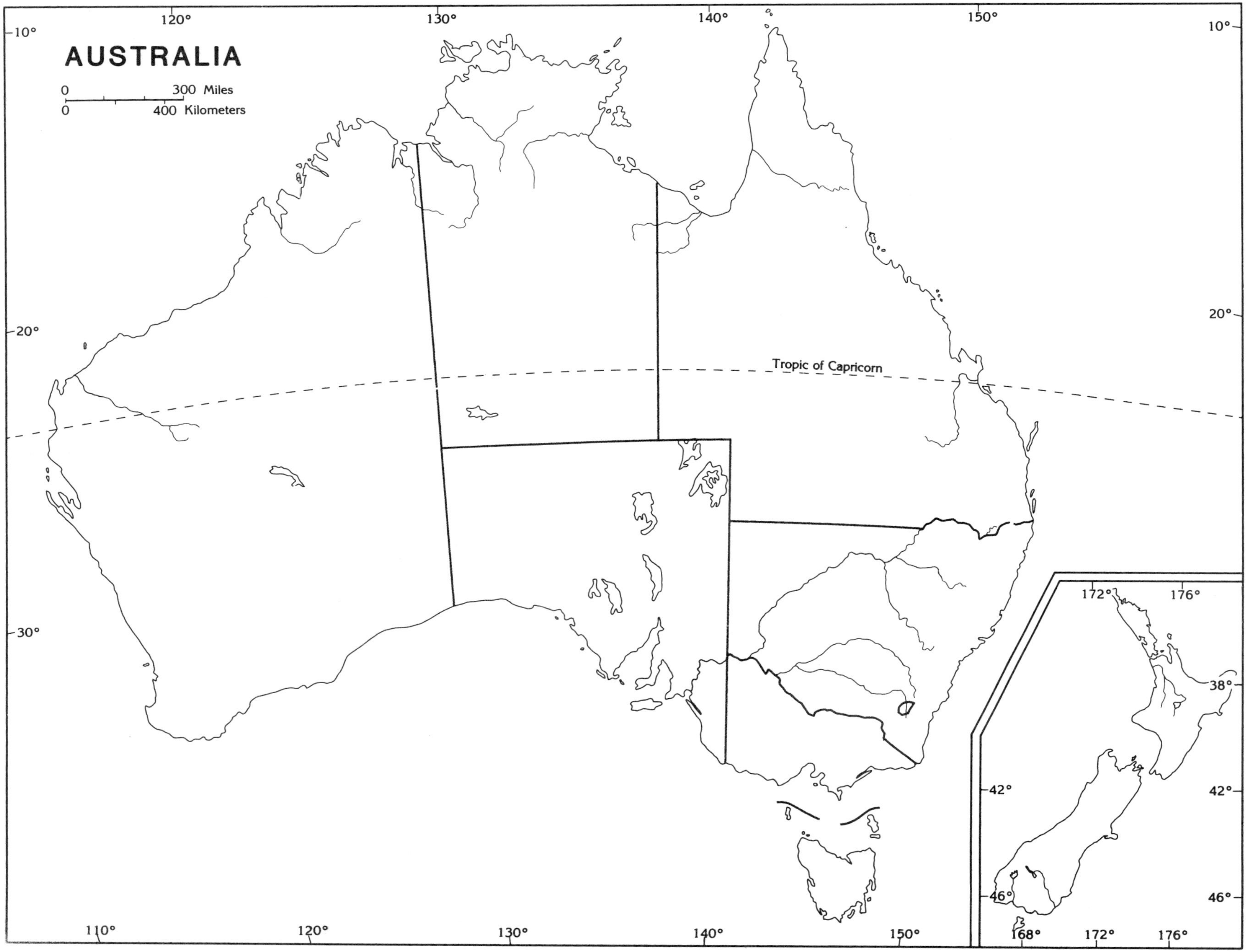

AUSTRALIA
0 300 Miles
0 400 Kilometers
Tropic of Capricorn
10°
20°
30°
110°
120°
130°
140°
150°
38°
42°
46°
168°
172°
176°

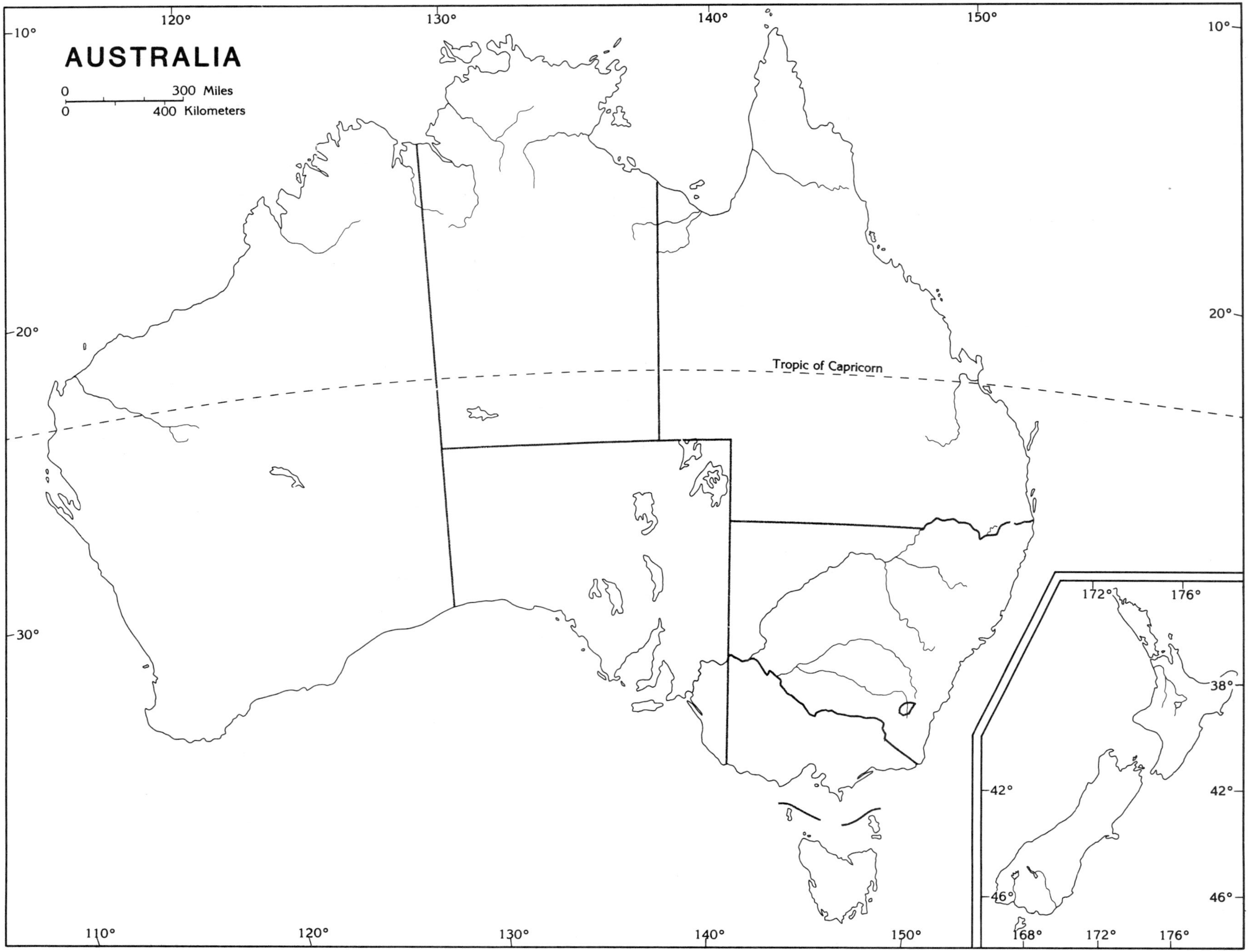
AUSTRALIA
0 300 Miles
0 400 Kilometers
Tropic of Capricorn
120°
130°
140°
150°
10°
20°
30°
110°
120°
130°
140°
150°
172°
176°
38°
42°
42°
46°
46°
168°
172°
176°

CHAPTER 2: THE SOVIET UNION--MOVING THE MONOLITH

OBJECTIVES OF THIS CHAPTER

Chapter 2 covers the Soviet Union, a world superpower and a key developed realm but with a socialist/communist economy and society. The Systematic Essay here--climatology--is appropriate for this vast territory that comprises the world's largest political unit in areal size, containing every climate type except the wet tropical category. After the world context of the U.S.S.R. is discussed, its historical cultural evolution is traced. This is followed by reviews of its physiographic, political, and ethnic frameworks. The centrally-planned Soviet economy is then discussed from a geographic perspective, highlighting agriculture and industry, and sets the stage for a survey for the realm's major economic regions. The chapter closes with an overview of geopolitical trends, population redistribution, and ongoing spatial change.

Having learned the regional geography of the Soviet Union, you should be able to:

1. Understand the basic workings of world climatology, the relationships between climate controls and elements, and the overall climatic pattern of the realm.

2. Grasp the essential ingredients of Russia's complex historical evolution, including the contributions of major groups that effected lasting change.

3. Understand the Soviet emergence as a global superpower, and its geographic resources in comparison to those of the United States.

4. Understand the political framework of the country's socialist republics and its relationship to the underlying cultural-geographical mosaic.

5. Understand the basic workings of the centrally-planned economy, and its relationships to the economic geographies of agriculture and manufacturing that continue to evolve.

6. Describe and map the leading functions of the realm's 7 major regions.

7. Explain Sir Halford Mackinder's heartland theory and its changing application to the world geopolitical scene.

8. Understand boundary and population trends in the contemporary Soviet state, particularly in the eastern reaches of the country.

9. Locate the major physical, cultural, and economic-spatial features of the realm on an outline map.

GLOSSARY

Climate (152)

A term used to convey a generalization of all the recorded weather observations over time at a certain place or in a given area; it represents an "average" of all the weather that occurs there. Climate may be regionalized into aggregates or groupings in which climatic elements--temperature, precipitation, wind, and pressure patterns--are similar. The climates of these regions are shaped by 8 interrelated processes or climatic controls: latitude, semipermanent pressure cells, wind belts, jet streams, ocean currents, altitude, mountain barriers, and land/water heating and cooling differentials.

Weather (152)

The state of the atmosphere at a location at any given moment.

Wind belts (152)

The primary circulation system of the atmosphere wherein 6 lateral belts emerge to exchange air among the semipermanent high and low pressure cells. In the ideal situation, moving poleward from the equator, one encounters the Northeast (0°-30°N) and Southeast Tradewinds (0°-30°S), Prevailing Westerlies (30°-60°N and S), and Polar Easterlies (60°-90° N and S).

Continentality (153)

Refers to degree of inland location away from the moderating effects of the oceans on the climates of adjacent landmasses. The most interior locations, such as the central U.S.S.R., experience both extreme heating and cooling because land surfaces heat up and cool off rapidly in the absence of moisture.

Monsoon (154)

Arabic word for "season"; refers to dry monsoon in winter when air flows offshore, and wet monsoon in summer when air flows onshore bringing huge quantities of moisture. Effect is result of land/sea pressure reversals by season, and is most pronounced in East, Southeast, and South Asia.

Gorod (154)

One of a network of fortified trading towns through which the Varangians organized the early economy of the Slavs in what is today the Ukraine.

Ostrogs (156)

Strategic way stations built by the Cossacks in river valleys across central and eastern Russia during their 16th and 17th century expansion from Muscovy, in the process defeating the Tatars and consolidating the Russian state all the way to the Pacific coast.

Soviet (156 box)

Revolutionary councils composed of workers' groups, formed after the 1905 revolution and vital as a grass-roots force in the ultimate success of the Bolshevik uprising in 1917.

Soviet Socialist Republic (163)

One of the 15 S.S.R.s that comprise the Soviet Union, each corresponding broadly to one of the country's major nationalities. Largest is the Russian Soviet Federated Socialist Republic (R.S.F.S.R.).

Acculturation (166)

Cultural modification resulting from intercultural borrowing. In cultural geography, the term is used to designate the change that occurs in the culture of indigenous peoples when contact is made with a technologically more advanced society.

Gosplan (166)

The State Planning Commission formed in the 1920s to oversee and implement the five-year plans whereby the U.S.S.R. has achieved significant economic progress. The major administrative body charged with centralized planning of the Soviet economy in line with Marxist-Leninist goals.

Sovkhoz (167)

Huge state farm, literally a grain- or meat-producing factory, making use of mass labor resources and large-scale agricultural mechanization.

Kolkhoz (167)

A smaller, local collective farm worked by peasants, devised as an intermediate step in the 1930s toward the achievement of a nationwide system of sovkhozes.

Analogue area analysis (168 box)

A method of comparative geographic study in which different resource-use systems are analyzed to see which is the more productive.

Povolzhye (176)

Russian name for the country of the middle and lower Volga Basin, used here as the Volga Region in the Soviet regionalization survey.

Donbas (177)

Contraction of Donets Basin, the Ukraine's and the U.S.S.R.'s leading coalfield and one of the realm's leading heavy-industry producers.

Virgin and Idle Lands project (178)

Ill-fated agricultural project of the 1950s that sought to expand wheat farming into the dry steppes of Kazakhstan; once again a growth pole, this time for industry as well as irrigated agriculture, centered by Tselinograd in the Karaganda Region.

Kuzbas (178-179)

Contraction of Kuznetsk Basin, the leading industrial zone of southern Siberia based on coal, iron ore, and good long-distance transport connections. Also forms sixth of the 7-region scheme used in text.

Heartland theory (181-182)

Geopolitical generalization derived by Sir Halford Mackinder, which saw pivot area or center of Eurasian continent as impregnable fortress again naval power--the supreme military force in 1904 when the concept was formulated. Pivot area later extended to Eastern Europe, perceived as gateway to the Eurasian heartland; popular with military planners of Nazi Germany, who followed such a strategy in their ultimately unsuccessful invasion of Russia in mid-World War II.

Rimland (182)

Counterargument centerpiece in alternative geopolitical scheme introduced by Nicholas Spykman, who claimed the rim or outer edge of Eurasia was far more important for potential world domination because of its far more considerable natural and population resources.

Frontier (182)

Predecessor of modern concept of political boundary, stemming from the time when the core areas of national states were separated by virtually empty zones.

Buffer state (184 box)

Remnant of bygone era in which European colonial powers carved out spheres of influence in the Third World. To avoid confrontation, certain territory was left relatively empty between such spheres as a cushion or "buffer." Mongolia has evolved as a buffer state between China and the U.S.S.R. Other Asian buffer states were Nepal, Thailand, and--until the 1979 Soviet invasion--Afghanistan.

SELF-TESTING QUESTIONS

Cover this side of the page with a sheet of paper. Uncover each line after you have attempted to answer the question in the left column. If necessary, refer to textbook page(s) listed at the right.

Question	Answer	Page
CLIMATOLOGY	CLIMATOLOGY	
What is the difference between climate and weather?	Weather is the state of the atmosphere at any moment in time; climate is the average weather over a long period of time.	152
Name the climatic elements.	Temperature, precipitation, pressure, and winds.	152

Name the eight climatic controls.	Latitude, semipermanent highs and lows, wind belts, jet streams, ocean currents, altitude, mountain barriers, and land/water heating differences.	152
Which major Koppen climate type is lacking in the Soviet Union?	Tropical (A) climates; all the other major types are present.	152
Why is continentality so important in Soviet climatology?	Because of the huge Eurasian landmass which experiences very hot summers and very cold winters.	153
What is the difference between the dry and wet monsoon?	The dry monsoon is the offshore airflow during the winter, whereas the wet monsoon is the onshore flow of rain-bearing winds during the hot summer.	154
SOVIET UNION CHARACTERISTICS	SOVIET UNION CHARACTERISTICS	
What is the traditional boundary between European Russia and Siberia?	The Ural Mountains.	151
When was the Soviet Union formed?	In 1917, as a result of the Bolshevik Revolution which overthrew the tsarist monarchy.	154
RUSSIAN HISTORY	RUSSIAN HISTORY	
Where was the first Slavic state located?	In the present-day Ukraine, northwest of the Black Sea.	154–155
When did Moscow emerge as the center of the Russian state?	In the 16th century, when Ivan the Terrible began to expel the Tatar conquerors.	155–156
What was the contribution of the Cossacks to the expansion of modern Russia?	These seminomadic peoples drove out the Tatars from Siberia and reached the Pacific by 1800.	156
What were the contributions of Peter the Great and Catherine the Great to the expansion of the Russian empire?	Peter opened the Baltic frontier, built the capital of St. Petersburg, and developed ties with Europe; Catherine pursued the southern frontier, obtaining warm-water ports on the Black Sea.	156–157
How far did the Russians penetrate into North America?	They reached across Alaska, the western coast of Canada, and as far south as San Francisco Bay; the U.S. later purchased Alaska from Russia in 1867.	158 box

PHYSIOGRAPHY	PHYSIOGRAPHY	
Where is the Russian Plain located?	Across all of European Russia as far east as the Urals; the Moscow Basin lies at its heart.	159
What are the two chief rivers of the West Siberian Plain, the world's largest lowland?	The Ob and the Irtysh.	159-160
What is the major inland water body of the Eastern Highlands?	Lake Baykal.	160
SOVIET POLITICS	**SOVIET POLITICS**	
Where was the capital moved to following the 1917 Revolution?	From St. Petersburg (now Leningrad) back to Moscow.	161
Name the 15 Soviet Socialist Republics.	R.S.F.S.R., Ukraine, Belorussia, Estonia, Latvia, Lithuania, Moldavia, Georgia, Armenia, Azerbaydzhan, Kazakh, Turkmen, Uzbek, Kirghiz, Tadzhik.	163
SOVIET ECONOMY	**SOVIET ECONOMY**	
Which economic sector has achieved the most growth since 1917?	Soviet industry.	166
What is the difference between a kolkhoz and a sovkhoz?	The former is a local collective farm run by peasants; the sovkhoz is a large state farm, mechanized, and run as a food-producing factory.	167
What is the greatest obstacle to farming in Soviet Central Asia?	The shortage of water, which necessitates irrigation throughout most of this region.	168
SOVIET REGIONS	**SOVIET REGIONS**	
Where is the Soviet Core located?	Within a 425-mile (700-kilometer) radius of Moscow, the heart of which is commonly called the Central Industrial Region.	173-174
Which cities are known as the Soviet Detroit and the Soviet Manchester?	Gorkiy and Ivanovo.	175
What is Leningrad's industrial specialization?	High-skilled engineering and machine-building.	175-176

Question	Answer	Page
What are the major energy resources of the Povolzhye?	The Volga valley produces large quantities of petroleum and natural gas.	176
What are the leading coal- and iron-producing centers of the Ukraine?	The Donets Basin (Donbas) and Krivoy Rog.	177
Why was the Urals Region developed for industry before World War II?	For strategic purposes against invasion from the west, and to exploit rich metallic mineral resources.	178
What is the leading manufacturing region of Siberia?	The Kuznetsk Basin (Kuzbas), centered south of the city of Novosibirsk.	179
Which city is centrally located in the industrial corridor of the Ussuri-Amur Basin in the Soviet Far East?	Khabarovsk.	180
POLITICAL GEOGRAPHY	POLITICAL GEOGRAPHY	
What is the heartland theory?	Mackinder's hypothesis was that whoever controlled the impregnable interior of Eurasia could rule the world.	181-182
What was the rimland theory?	Spykman's hypothesis was that the outer edges (rimland) of Eurasia contained a greater power potential.	182
Why is Mongolia a buffer state?	It has functioned as a cushionlike frontier zone between Russia and China.	184 box
U.S.S.R. IN THE 1980s	U.S.S.R. IN THE 1980s	
Why has the Soviet birth rate been quite low since 1917?	Tens of millions of lives, and therefore unborn children, were lost in two devastating world wars.	185
Why has the Soviet urbanization rate lagged behind Europe's?	Overcrowded cities are unable to house millions who would prefer to migrate to urban areas.	185
What is the BAM project?	The high-speed Baykal-Amur Mainline railroad will open up a new corridor across the Far Eastern Region, and will also shorten the rail route across this area.	186

MAP EXERCISES

MAP COMPARISON

1. The map of Soviet climates (Fig. 2-2, p. 153) represents one pattern of regional distribution for a large landmass lying between approximately 30° and 70°N. Compare this pattern to the one exhibited by North America (Fig. I-10, p. 18), noting similarities and differences; also, compare the Soviet and North American maps to the Hypothetical Continent pattern (Fig. I-11, p. 20) and note which real-world pattern is a "better fit" to the model (and why).

2. The Soviet Union's republics contain a sizable non-Russian population as Fig. 2-10 (p. 165) reveals. Compare the distribution of Soviet ethnic groups (Fig. 2-8, p. 164) with the map of Soviet republics, and compare both maps to the population distribution of the U.S.S.R. (Fig. I-14, p. 27), noting those areas where the Russians are most underrepresented--this should give you a insight into the internal spatial patterns of the republics that are mapped in Fig. 2-10.

3. How do the Soviet agricultural regions (Fig. 2-12, p. 167) and railway routes (Fig. 2-14, p. 170) compare to the national population distribution (Fig. I-14, p. 27)? Are there distinctive differences on opposite sides of the Ural Mountains? If there are maladjustments, are current population redistribution trends helping to balance the situation?

4. Compare the maps of Russian (Fig. 2-3, p. 157) and American historical expansion (Fig. 3-10, p. 208). What similarities and differences were faced by each country, and did their experiences in the 19th century affect their mutual attitudes toward each other in the 20th?

5. Briefly compare all of the maps in Chapter 2 with those of Chapter 3 on North America. Make observations on the geographical "balance of power," and list three examples of one superpower having an advantage over the other.

MAP CONSTRUCTION

(Use outline maps at the end of this chapter)

1. In order to familiarize yourself with Soviet physical geography, place the following on the first outline map:

   a. Rivers: Dnieper, Don, Volga, Dvina, Ural, Kama, Ob, Irtysh, Yenisey, Syr Darya, Amu Darya, Angara, Lena, Aldan, Kolyma, Amur, Ussuri, Tunguska

   b. Water bodies: Caspian Sea, Black Sea, Pripyat Marsh, Sea of Azov, Aral Sea, Lake Baykal, Lake Balkhash, Sea of Okhotsk, Bering Sea, Barents Sea, White Sea, Baltic Sea

   c. Land bodies: Crimean Peninsula, Kola Peninsula, Novaya Zemlya, Kamchatka Peninsula, Kurile Islands, Sakhalin, Kirghiz Steppe, Moscow Basin, West Siberian Plain, Yakutsk Basin

d. Mountains: Ural Mountains, Caucasus Mountains, Verkhoyansk Mountains, Altai Mountains, Pamir Mountains, Sikhote Alin Mountains, Stanovoy Mountains

2. On the second map, political-cultural information should be entered as follows:

a. Draw in each S.S.R. boundary and label each republic

b. Reproduce the ethnic map (Fig. 2-8, p. 164) in general form using an appropriate set of area symbols (preferably color pencils)

3. On the third outline map, economic-urban information should be entered as follows:

a. Cities (locate and label with the symbol • /label Moscow and all S.S.R. capitals with the symbol *): Moscow, Leningrad, Tallinn, Riga, Vilnius, Kaliningrad, Minsk, Kiev, Lvov, Kishinev, Odessa, Donetsk, Rostov, Krivoy Rog, Dnepropetrovsk, Kharkhov, Voronezh, Kursk, Tula, Gorkiy, Ivanovo, Yaroslavl, Murmansk, Arkhangelsk, Perm, Kazan, Ufa, Kuybyshev, Saratov, Volgograd, Astrakhan, Groznyy, Tbilisi, Baku, Yerevan, Nizhniy Tagil, Sverdlovsk, Chelyabinsk, Magnitogorsk, Askhabad, Tashkent, Dushanbe, Frunze, Alma-Ata, Karaganda, Tselinograd, Omsk, Novosibirsk, Krasnoyarsk, Bratsk, Irkutsk, Yakutsk, Khabarovsk, Vladivostok, Komsomolsk, Magadan

b. Economic regions (identify with circled letter):

A - Minsk Oblast (Central Belorussia)
B - Central Industrial Region
C - Povolzhye
D - Ukraine Industrial Region
E - Donbas
F - Urals Industrial Region
G - Karaganda/Tselinograd Region
H - Kuzbas
I - Yakutia
J - Soviet Central Asia

PRACTICE EXAMINATION

SHORT-ANSWER QUESTIONS

Multiple Choice

1. The political geographer who first proposed the heartland theory was:

a) Ratzel
b) Spykman
c) Koppen
d) Mackinder
e) Peter the Great

2. What was the capital of the Soviet Union in the early years following the 1917 Bolshevik Revolution:

a) Moscow
b) Leningrad
c) Petrograd
d) Kiev
e) St. Petersburg

3. Novosibirsk is a leading manufacturing center in the Soviet region called:

a) European Russia
b) the Kuzbas
c) the Baltic republics
d) Central Asia
e) Karaganda

True-False

1. The Prevailing Westerlies are wind belts lying in both the eastern and western hemispheres between 30° and 60° longitude.

2. The Soviet Union is the world's largest state in population size.

3. Krivoy Rog and the Donets Basin are both located in the Ukraine Industrial Region.

Fill Ins

1. The large buffer state of far eastern Eurasia lying between portions of the U.S.S.R. and China is ________________.

2. The ________________ River is most closely associated with the region known as the Povolzhye.

3. The American state named ________________ was originally purchased from Russia in the 1860s.

Matching Questions on Soviet Cities

_____	1.	Leading Ukraine iron-mining center	A.	Leningrad
_____	2.	Oil center on Caspian Sea	B.	Irkutsk
_____	3.	Formerly St. Petersburg	C.	Novosibirsk
_____	4.	Pacific terminus of Trans-Siberian Railroad	D.	Krivoy Rog
_____	5.	118°F (66°C) annual temperature range	E.	Ivanovo
_____	6.	Soviet Pittsburgh	F.	Baku
_____	7.	Lake Baykal	G.	Vladivostok
_____	8.	Kuznetsk Basin	H.	Verkhoyansk
_____	9.	Soviet Manchester	I.	Tselinograd
_____	10.	Virgin and Idle Lands project	J.	Donetsk

ESSAY QUESTIONS

1. It is claimed that the modern Soviet empire, to a large extent, is a legacy of St. Petersburg and European Russia--not the product of Moscow and the Communist Revolution. In this context, discuss the course of Russian history and territorial evolution in the two centuries prior to 1917, and conclude your overview with a statement as to whether you concur or not with the above interpretation.

2. The Soviet experiment with a centrally-planned economy for the past seven decades has been one of the U.S.S.R.'s notable organizational enterprises. Discuss the progress--or lack thereof--that central planning has achieved, contrast the varying experiences of the manufacturing and agricultural sectors, and speculate where the country might today be economically had it chosen a different control structure.

3. It is often said that communism is an effective system--for Russia. Discuss the overall gains that this kind of economic system has brought to the average citizen, the adaptability of Russian society to communism, and how the country might change if a capitalist system was introduced.

4. More than most national capitals and primate cities, Moscow is the critical focal point for a state, a world superpower, an economic colossus, and a sprawling political territory that is the largest on earth. Discuss the historical, cultural, physical, and economic advantages that Moscow possessed to allow it to take such effective control of the U.S.S.R.; also, present a synopsis of the Central Industrial Region which Moscow anchors, reviewing the productive activities that have concentrated there and further enhanced the geographic position of the capital.

5. As with the American west, the Soviet push toward the east has been propelled by similar desires to exploit rich resources and bring remote areas into more effective national control. Discuss the progress of the eastward push during tsarist times, trace the Soviet effort to develop Siberia since 1917, and present an overview of this vast area's opportunities that are now challenging the Soviets to undertake their mightiest efforts yet to open up this last frontier.

TERM PAPER POINTERS

The "Term Paper Pointers" section of Chapter 1 in this Study Guide offered suggestions about approaching research and writing on geographic realms and their components, and you may wish to consult this material if you are undertaking a report on a Soviet region.

Because of its gargantuan size and complexity as well as its less-Westernized culture, it is especially important to get hold of a good regional geography that penetrates more deeply into the subject than we are able to in this textbook. Fortunately, several good books are easily available; among the recommended titles cited on text pp. 188-189 are Cole, Dewdney [1979], Gregory, Hooson [1966], Howe, Lydolph [1977], Mellor--and particularly Parker and Symons et al. [both 1983], which are quite up-to-date (currency of data is always a problem in the case of the U.S.S.R.). It is also worthwhile to seek additional perspectives. Good sources of various maps are Chew and Dewdney [1982], and compendia of current information are Brown, Katz, and Scherer. Historical dimensions, so important in the interpretation of current events, can be found in Bater & French, Hopkirk, Linz, and Pipes; equally significant perspectives on the economic system are found in Nove and Bergson & Levine; social/ethnic considerations are treated in Allworth, Clem [1980], and Karklins, Ro'i, and Wixman and even the "Inside the U.S.S.R." issue of *Time* [June 23, 1980] is useful.

Several *topical* treatments of Soviet geography also exist. Among the politico-geographical works are Armstrong et al., Bialer, Glassner & de Blij, Hooson [1962], both Mackinder selections, Parker [1982], Pepper & Jenkins, Prescott, and Spykman. Urban geography is covered in C. Harris, French & Hamilton, Giese, Andrusz, Bater, Morton & Stuart, and Hamilton. Economic geography is particularly well represented: the resource base in Jensen et al., Jackson, Dienes & Shabad, Shabad [1969], and Shabad & Mote; transportation in Hunter, Ambler, Mote, North, Sanchez [1986], and Symons & White; agriculture in Hedlund, Duetsch, R. Stuart, and Symons; industry in Dewdney [1976], Bradshaw & North, and Zum Brunnen & Osleeb; energy in Sagers & Green, *U.S.S.R. Energy Atlas*, and Wilson. Finally, it is often fascinating to see how a realm's own scholars view *their* countries-- Demko & Fuchs [1974] provide insights, and recent works from Moscow are Mikhailov and Spidchenko.

The *climatology* Systematic Essay may also be used as a springboard for a research paper. Fortunately, several fine, non-technical books are available, among them Critchfield, Fairbridge & Oliver, Gedzelman, Greenland & de Blij, Hare, Lydolph [1985], and Mather (the last of special interest because of its coverage of applied climatology).

Climatic data are especially useful as well, either as supporting information for charts, tables, and maps, or even for developing an entire paper around. For the United States, a great wealth of data and supporting studies can be found in Climate and Man (cited on p. 188)--even though this 1941 Yearbook of Agriculture is almost a half-century old, it is available in libraries and is considered a great classic by American geographers and climatologists. For the rest of the world, the references in the books listed above should steer you in the right direction.

SPECIAL EXERCISE

Comparisons of the Soviet Union and the United States are constantly in the news as the two superpowers try to maintain a rough strategic balance in the world. Geographical comparisons can be just as enlightening, and this special exercise focuses on such an evaluation (n.b.: two extra outline maps of the U.S.S.R. and North America are attached for this purpose).

The last paragraph on page 158 in the text provides the cue for a comparison of the natural environments of the superpowers. As the book prompts, locate the comparable soil zones in the Soviet Union and North America; on the first pair of outline maps, plot soil and precipitation working from Figs. I-13 (pp. 14-15) and I-9 (pp. 14-15). Record your observations about the environmental differences you see, and discuss how the less fortunate Soviet situation may have contributed to the country's persistent agricultural problems that were discussed in the text.

On the second set of outline maps, prepare a comparison of American (including Canadian) and Soviet strategic natural resources. This can best be compiled from an up-to-date world atlas, and you may want to choose the most useful from among the atlases held in your library's reference department. For both the Soviet Union and North America, map the distributions of the following resources (each in a different color): petroleum, natural gas, coal (anthracite, bituminous or sub-bituminous), iron ore, nickel, uranium, copper, bauxite (aluminum), manganese, and lead. Once completed, present a comparative overview of the two maps, identifying the relative strengths and weaknesses you observe in each superpower as measured against the other. A helpful source for the U.S.S.R. beyond the sources suggested above is Jensen (cited on p. 188), a comprehensive work that is based on some of the newest data available in the West. The same is true for the U.S.S.R. Energy Atlas, a widely available CIA publication that contains a splendid overview of Soviet energy potentials.

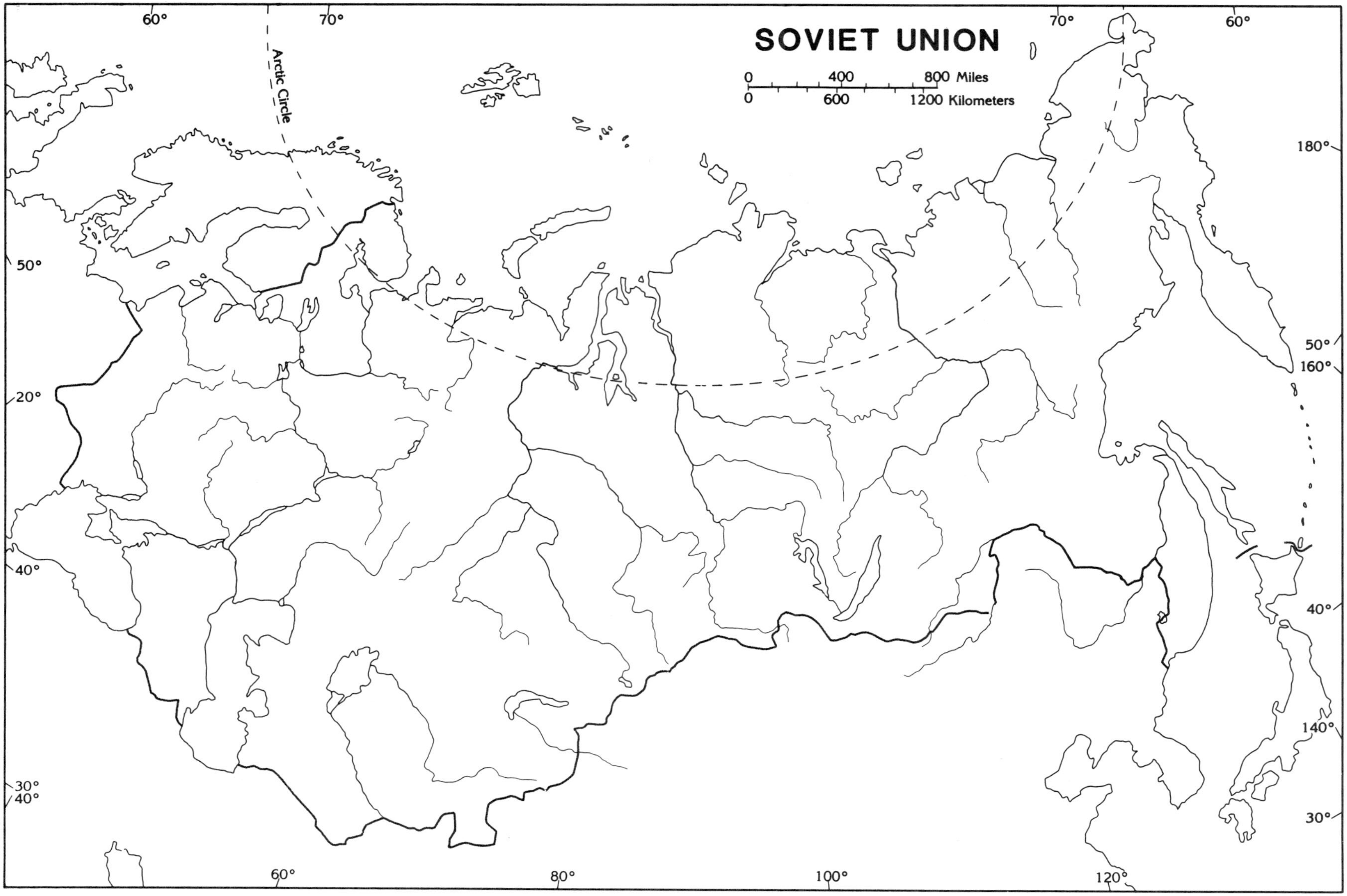
SOVIET UNION
0 400 800 Miles
0 600 1200 Kilometers
Arctic Circle
60°
70°
70°
60°
50°
20°
40°
30°
40°
180°
50°
160°
40°
140°
30°
60°
80°
100°
120°

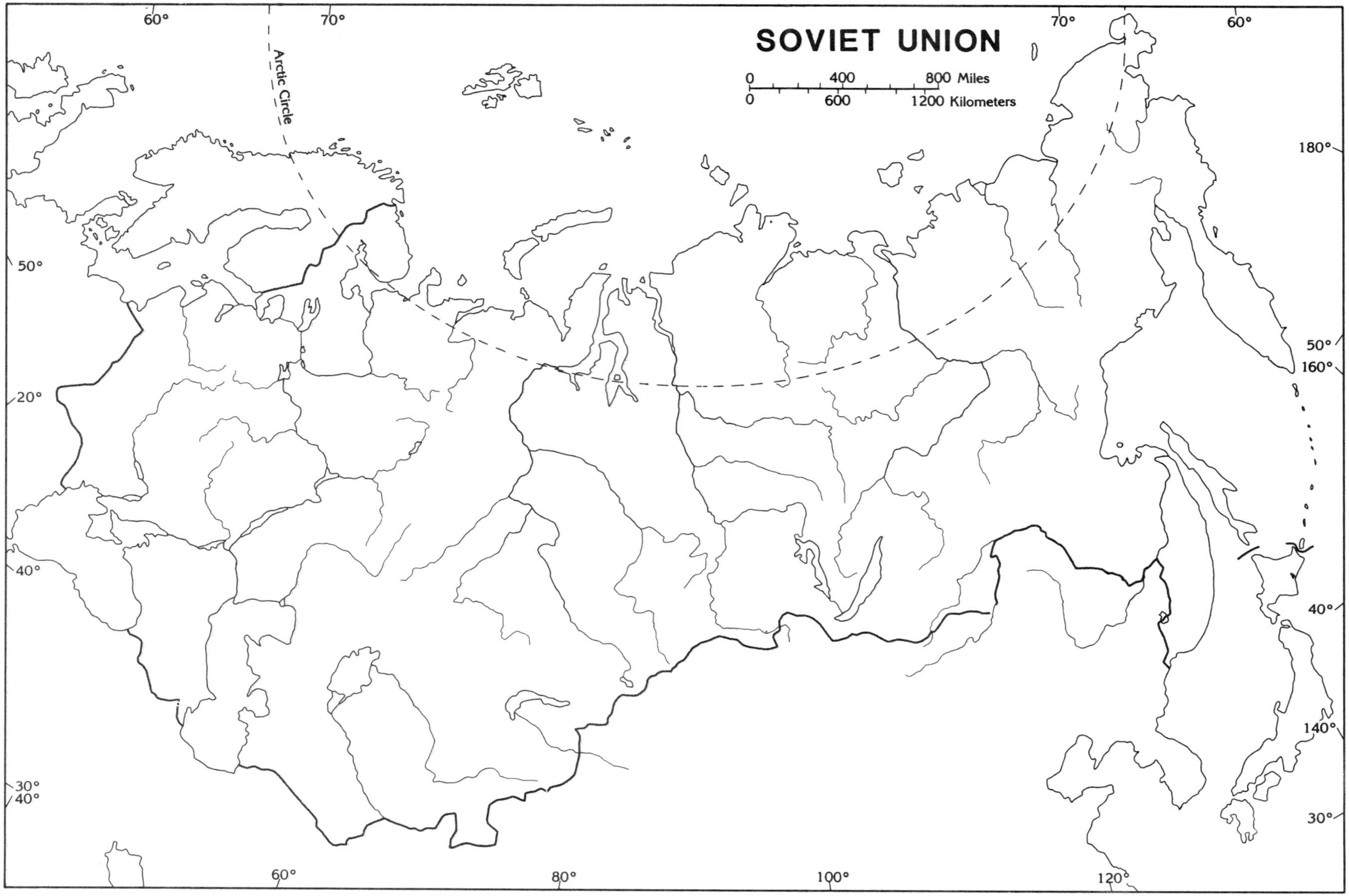
SOVIET UNION
0 400 800 Miles
0 600 1200 Kilometers
Arctic Circle
60°
70°
70°
60°
180°
50°
160°
50°
20°
40°
40°
140°
30°
30°
40°
60°
80°
100°
120°

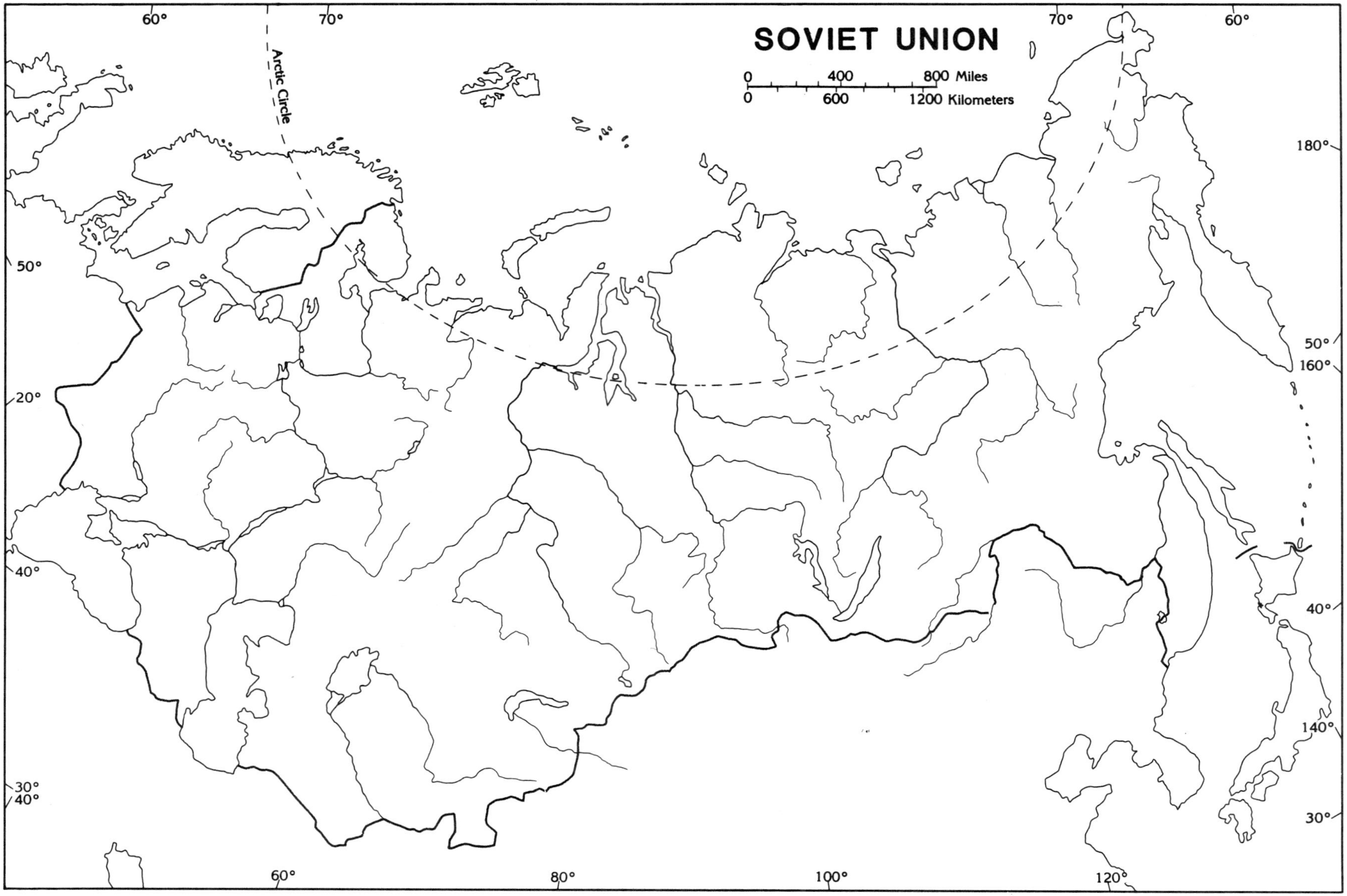
SOVIET UNION
0 400 800 Miles
0 600 1200 Kilometers
Arctic Circle
60°
70°
70°
60°
180°
50°
160°
40°
140°
30°
50°
20°
40°
30°
40°
60°
80°
100°
120°

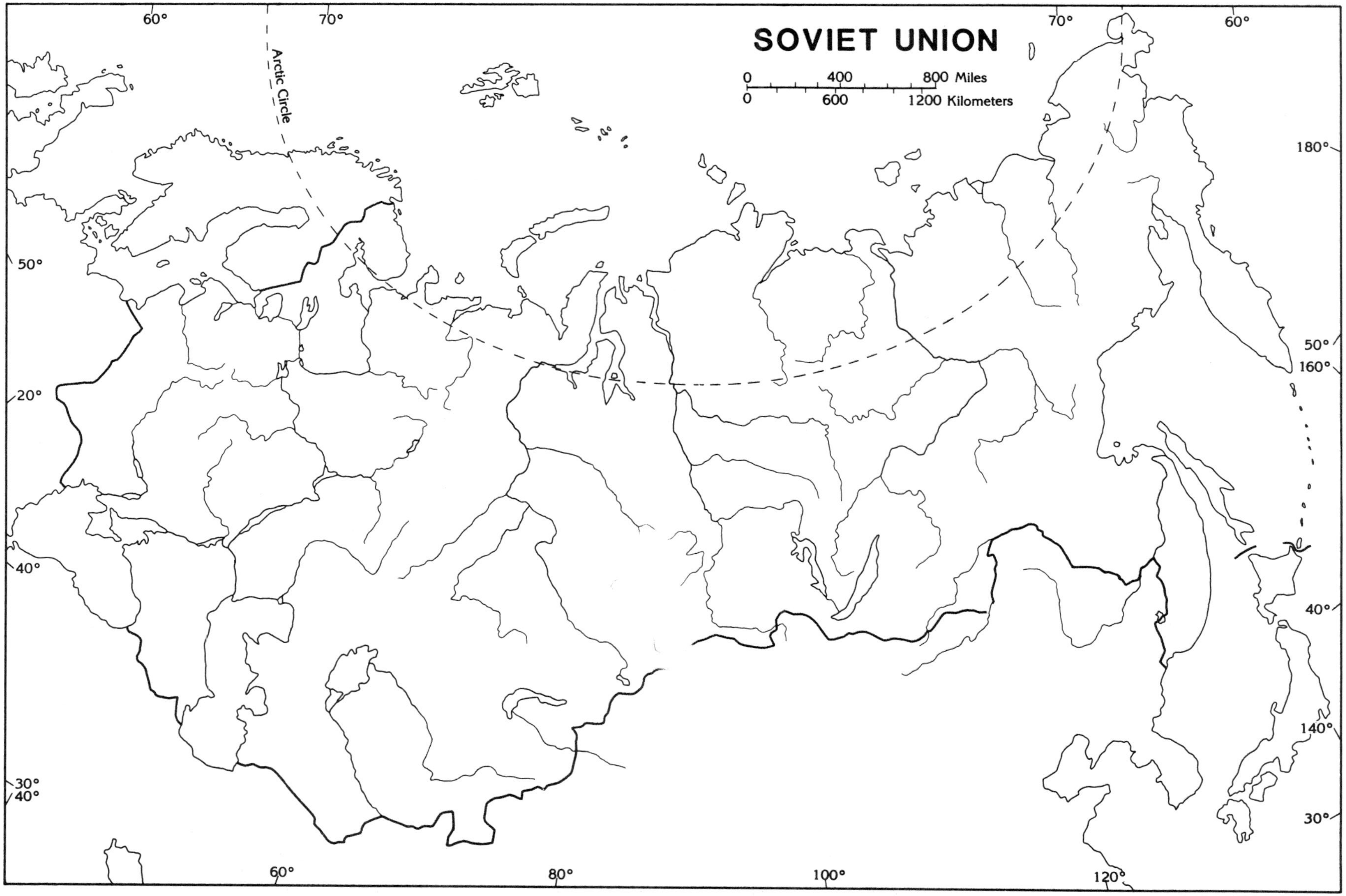
SOVIET UNION
0 400 800 Miles
0 600 1200 Kilometers
Arctic Circle
60°
70°
70°
60°
180°
50°
160°
40°
140°
30°
50°
20°
40°
30°
40°
60°
80°
100°
120°

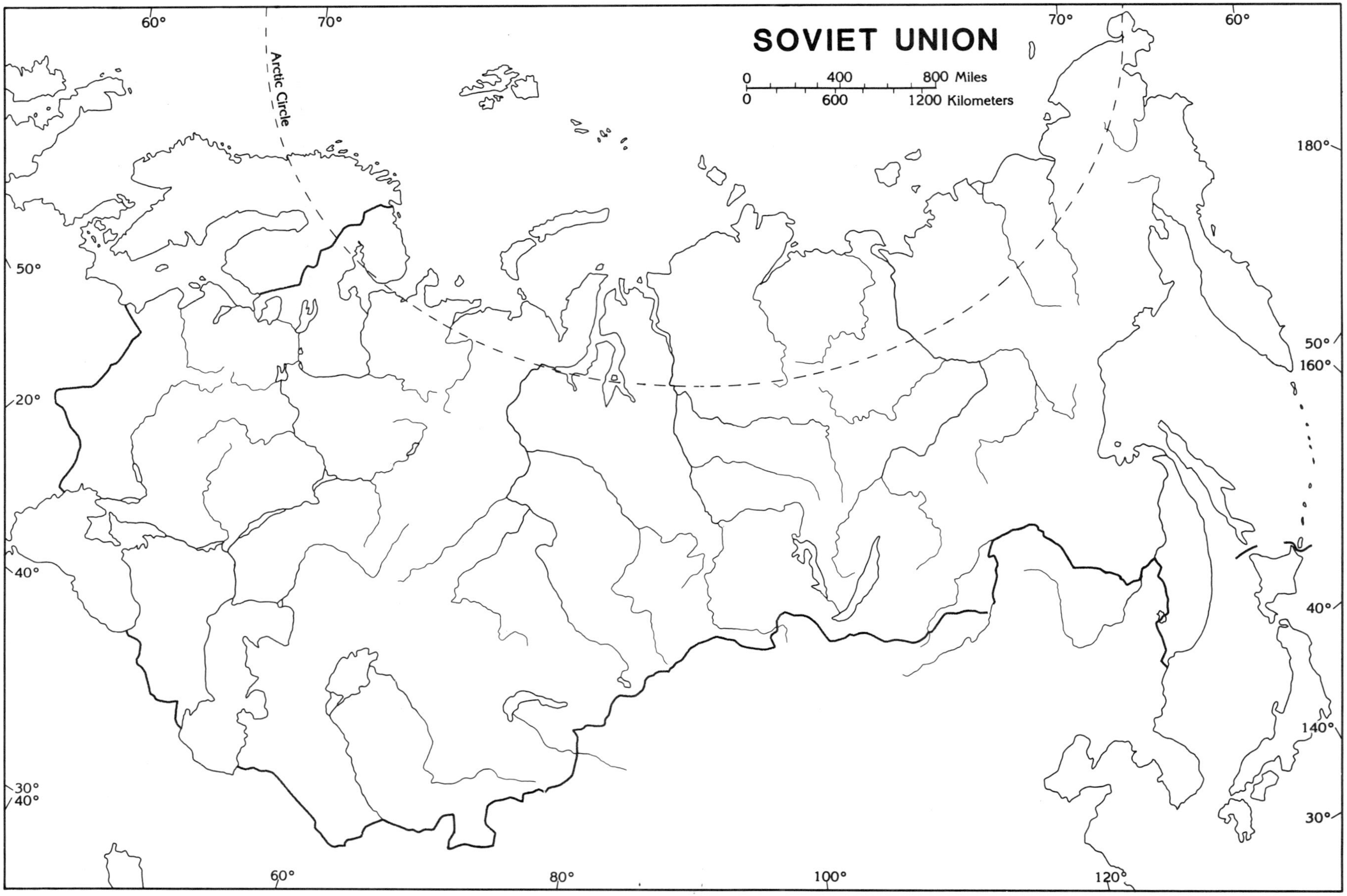
SOVIET UNION
0 400 800 Miles
0 600 1200 Kilometers
Arctic Circle
60°
70°
70°
60°
180°
50°
160°
50°
20°
40°
40°
140°
30°
40°
30°
60°
80°
100°
120°

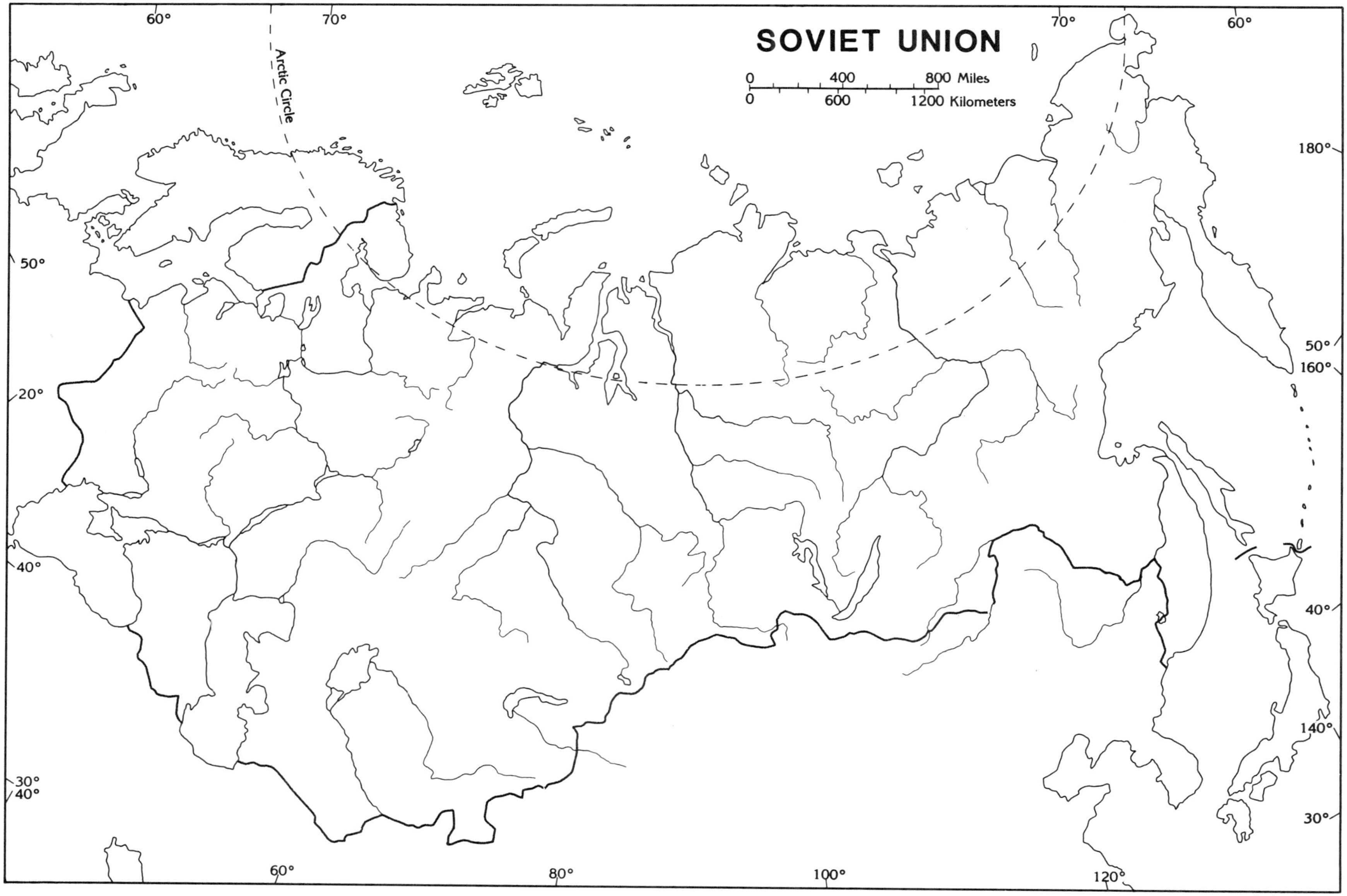
SOVIET UNION
0 400 800 Miles
0 600 1200 Kilometers
Arctic Circle
60°
70°
70°
60°
180°
50°
160°
40°
140°
30°
50°
20°
40°
30°
40°
60°
80°
100°
120°

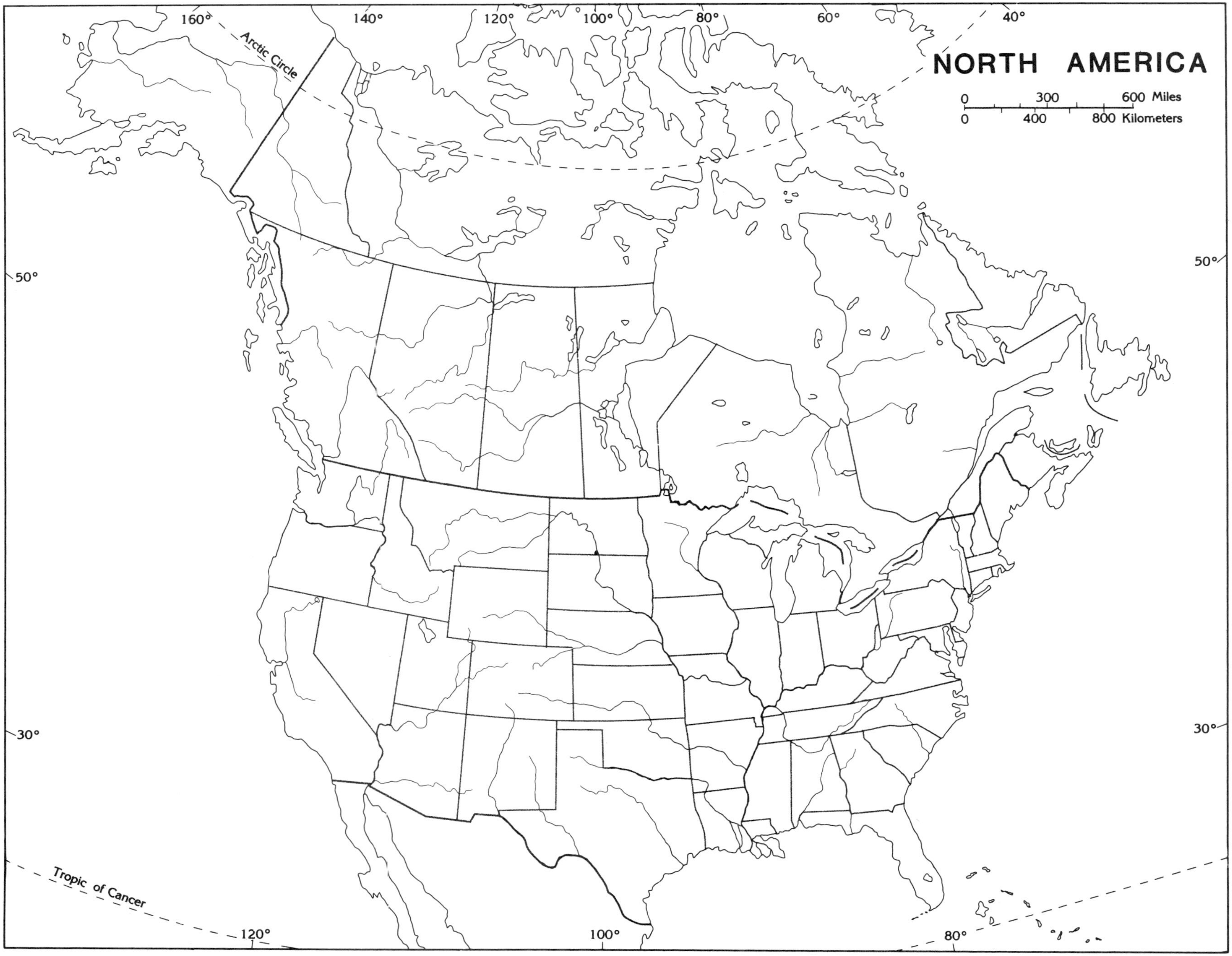
NORTH AMERICA
0 300 600 Miles
0 400 800 Kilometers
160°
140°
120°
100°
80°
60°
40°
50°
30°
Arctic Circle
Tropic of Cancer
120°
100°
80°

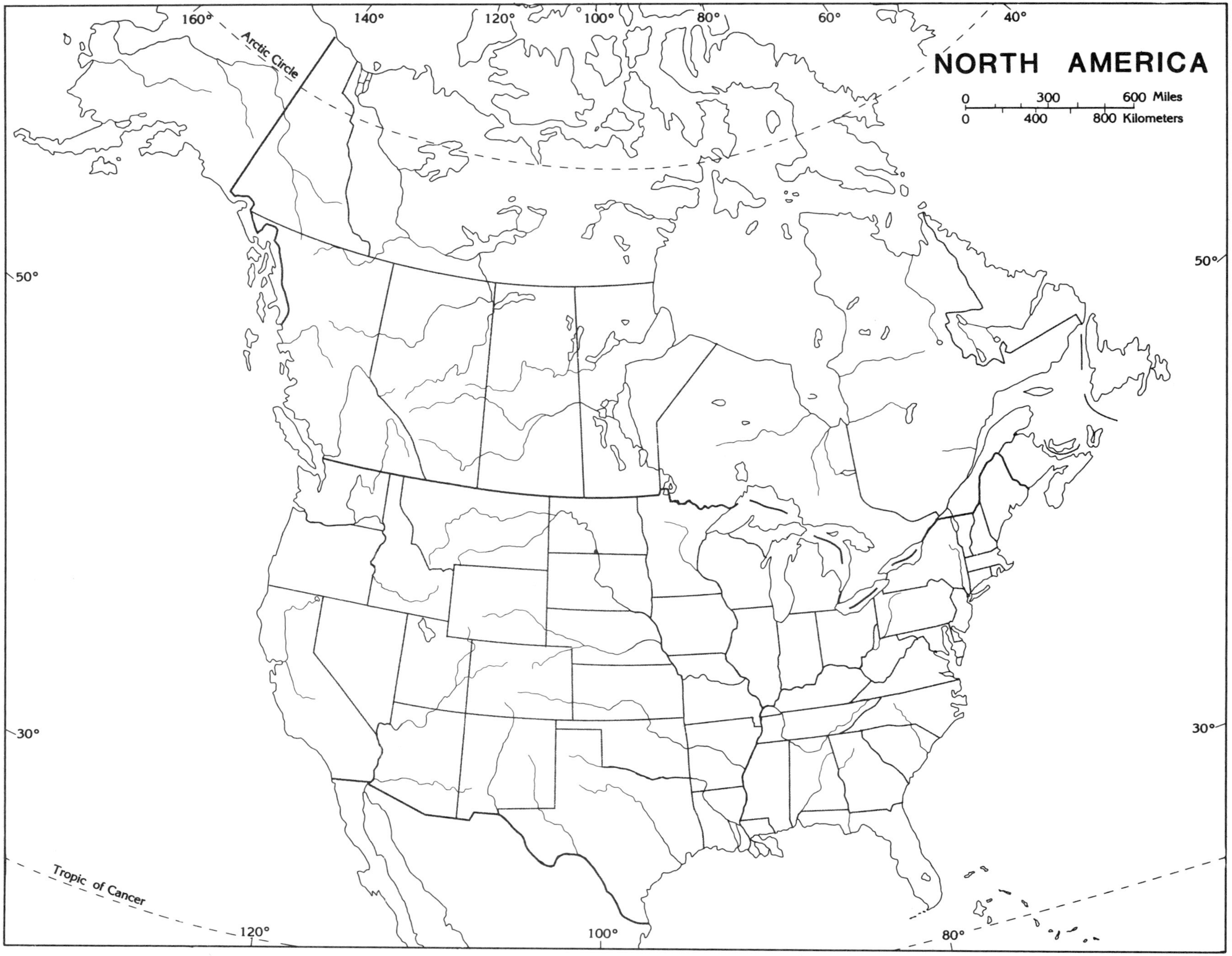
NORTH AMERICA
0 300 600 Miles
0 400 800 Kilometers
Arctic Circle
Tropic of Cancer
160°
140°
120°
100°
80°
60°
40°
50°
30°
120°
100°
80°

CHAPTER 3: NORTH AMERICA--THE POSTINDUSTRIAL TRANSFORMATION

OBJECTIVES OF THIS CHAPTER

Chapter 3 covers that part of the world you know best, but rapid change persists in the U.S. and Canada as this century draws to a close. As North America's societies and economies complete their postindustrial transformation, a new geographic reality is emerging. Although this transition is still going on, many of its broad features have become evident since 1970 and they form the foundation for this re-examination of the realm's regional human geography. The hypotheses of the "galactic metropolis" and the "nine nations," the new dominance of urban realms and the outer suburban city on the metropolitan scene, the economic-geographic significance of the Silicon Valleys and other lineaments of postindustrial society--these are just some of the developments that are reshaping America which are of such recent vintage that they were still years in the future when the first edition of the book appeared in 1971.

The chapter opens with a brief survey of urban geography, an appropriate choice for a Systematic Essay to accompany the most advanced and urbanized of all the world's realms. After the background introduction to the contemporary United States and Canada--which focuses on the unique composition of its populations--the continent's natural environment is reviewed. Population is then treated to a historical spatial analysis, emphasizing settlement patterns in both rural and urban contexts; the latter metropolitan dimension is examined at some length, covering both macro-scale and internal urban-growth patterns well into the 1980s. Cultural geography follows, emphasizing traditional approaches as well as more recent breakthroughs in the study of environmental perception and spatial behavior. We then turn to economic geography, stressing the emerging postindustrial sectors and re-examining agriculture and manufacturing from this new perspective. A concluding regional survey integrates many of the chapter's themes, and includes a comparison to Joel Garreau's imaginative nine-nations scheme.

Having learned the regional geography of North America, you should be able to:

1. Describe the main concerns of urban geography, and the kinds of research and practical-application questions that its proponents pursue.
2. Understand the similarities and differences of the U.S. and Canadian populations.
3. Grasp the essentials of the physical geography of this realm, including the intensifying problems of human environmental impacts.
4. Trace the historical geography of settlement across rural America in the 19th century, and within urban America since 1900.
5. Understand the main processes and patterns of urban growth, at both the scales of the national central-place hierarchy and the internal structuring of the individual metropolis.
6. Grasp the ongoing changes of the 1980s, as postindustrialism transforms culture and the economy, and urban and regional spatial change intensify.

7. Understand contemporary American economic geography--the shifting resource base, energy patterns, the historical Thunian framework of agricultural regionalization, the dilemmas of declining smokestack industry, and the postindustrial revolution's far-reaching spatial impacts.

8. The changing regional infrastructure of North America, that is provocatively brought into focus by the nine-nations scheme.

9. Locate the major physical, cultural, and economic-spatial features of the realm on an outline map.

GLOSSARY

Interurban geography (192)

The macro-scale approach that treats urban systems in large areas, with individual cities viewed as points in a hierarchical network that interact with one another and serve hinterlands of appropriate sizes.

Intraurban geography (192)

Focuses on the internal spatial organization of the individual metropolis, emphasizing structural form as well as the distribution of people and activities in central cities and suburbs.

Central Place Theory (192, 212-213)

Location theory devised by Walter Christaller and his followers; it seeks to explain the spatial distribution of urban places of varying size and function.

Concentric-zone model (192)

A geographical model of the American central city that suggests the existence of five concentric rings, (that is, arranged around a common center).

Sector model (192)

An alternative model of the American central city, devised by Homer Hoyt, which argued that radial transport routes were so important that they shaped an intracity structure dominated by sectors emanating outward from downtown.

Multiplenuclei model (192-193)

The Harris-Ullman model that showed the mid-twentieth-century American central city to consist of several zones arranged around nuclear growth points.

"Rust Bowl" (191)

Popular term used to describe the aging smokestack-industry facilities of the northern and eastern United States.

"Nine nations" scheme (195, 244-245)

Revisionist regionalization scheme proposed by Washington Post editor Joel Garreau in his 1981 book The Nine Nations of North America--see text pp. 244-245 for an overview.

Fragmented state (195)

A discontinuous country whose territory consists of two or more separated areal political units; the United States is such a state, with Alaska and Hawaii occupying space separated from the conterminous 48 "mainland" states.

Francophone (197)

Country or region in which several languages are spoken, but where French is the lingua franca or language of the elite; Quebec is Francophone Canada.

Pluralistic Society (197)

A society composed of multiple social groups. In Canada, divisions run along ethnic and linguistic lines; in the U.S., these divisions have largely been obliterated by the "melting pot" that absorbed the upwardly mobile majority--however, the major social cleavage has occurred along racial lines which still fosters widespread residential segregation.

Residential segregation (197)

The refusal of whites to share their immediate living space with nonwhites across most of the U.S. results in the concentration of whites and blacks into separate, racially-distinct neighborhoods; for blacks this is an involuntary process as dualistic housing-market mechanisms steer them away from white areas--thus, "integration" (in the words of 1960s activist Saul Alinsky) is too often the short time between the arrival of the first blacks and the departure of the last whites.

Time-space convergence (198)

Conceptualization of the idea that the world is shrinking--i.e., that breakthroughs in movement technology over time literally bring places ever nearer to each other in travel time, thereby facilitating the wider dispersal of people and activities and constant spatial reorganization.

Physiographic province (198)

The clear and well-defined divisioning of a continental-scale land surface into physically uniform regions, marked by a general homogeneity in relief, climate, vegetation, soils, and other environmental variables.

Orographic precipitation (200 box)

Mountain-induced precipitation, especially where airmasses are forced over topographic barriers. Areas beyond such a mountain range experience the rain shadow effect.

Rain shadow effect (200-201 box)

The relative dryness in areas downwind of, or beyond, mountain ranges caused by orographic precipitation, whereby moist airmasses are forced to deposit most of their water content.

Arid and Humid America (200, 201 box)

The broad divisioning of natural environments in the United States, with the wide boundary following, more or less, the line along which 20 inches (50 centimeters) of rainfall are received annually. The average position of this north-south line is approximately 100°W longitude, located in the central Great Plains; to the east is Humid America and to the west lies Arid America (except for the narrow moist zone between the Pacific coast and its nearby parallel ranges. Humid America is generally associated with natural forest vegetation and acidic soils, whereas Arid America contains steppe or short-grass prairie vegetation and alkalinic soils.

Isohyet (199)

A line connecting all places receiving the same amount of annual precipitation, such as the 20-inch (50-cm.) isohyet that divides Arid and Humid America (see above). An isoline, of which an isohyet is one specific example, is a line connecting all points recording the same value of a phenomenon (e.g., temperature, elevation, air pressure, and the like).

Chaparral (201)

The distinctive brushy tree-and-shrub landscape typical of Mediterranean climates (Koppen Cs), called chaparral in Southern California and maquis in Mediterranean Europe.

Aquifer (203)

An underground reservoir of water contained within a porous, water-bearing rock layer. This rock stratum is often overlain by impervious layers to prevent upward seepage of groundwater by capillary action.

Smog (203)

Contraction of the words "smoke" and "fog," this is a severe form of artificial air pollution produced by industrial effluents and/or automobile-exhaust emissions. Usually occurs when a warm, dry layer of air well above the surface prevents cooler air below from rising, causing a "greenhouse" effect that traps air which becomes progressively more stagnant and polluted.

Sunbelt (205)

The so-called "southern tier" of the conterminous United States that increasingly grows in population and economic activity at the expense of the rest of the country (the so-called "Frostbelt" or "Snowbelt"). It must be stressed these are popularly used terms that do not stand up to the more rigorous geographer's definitions of a region; the northern boundary of the Sunbelt is also in dispute--most commonly, it follows 37°N latitude.

Culture hearth (207)

Heartland, source area, place of origin of a major culture.

Township-and-Range land division system (209)

Checkerboardlike land division system, devised by Thomas Jefferson, to easily survey property lines to accommodate farmers settling west of the Appalachians (except in Texas); leaves highly distinct geometry in the rural landscape.

Urban hierarchy (210, 213 box)

The rank ordering of central places in an urban system with the primate city at the top, followed in descending order by other cities, towns, villages, and hamlets; each level in the hierarchy is subordinate to the one above it and superior to the one below, as measured by the range of the highest-order good based in each center.

Borchert model of urban-system evolution (210-211)

A generalization of the historical growth of the U.S. urban system according to key changes in transportation and industrial energy-use that occurred within four stages of development: the Sail-Wagon Epoch (1790-1830), the Iron Horse Epoch (1830-70), the Steel-Rail Epoch (1870-1920), and the Auto-Air-Amenity Epoch (1920-1970).

Core area of North America (211)

As mapped in Fig. 3-12 (p. 211), the American Manufacturing Belt of the northern/eastern U.S. and southeastern Canada, cornered by Boston, Milwaukee, St. Louis, and Washington, and containing the dozen industrial districts listed on p. 214.

Megalopolis (214)

More specifically, the Atlantic Seaboard or Boston-Washington ("Boswash") Megalopolis, which contains the economic heartland of the U.S. and forms the trading "hinge" between the nation and the rest of the world.

Main Street (214)

The "Megalopolis of Canada," that country's primary conurbation stretching southwest from Quebec City through Montreal and Toronto to Windsor on the U.S. border across the river from Detroit.

Adams model of intrametropolitan evolution (215-216)

A generalization of the historical growth of the American metropolitan city according to key transport breakthroughs that occurred wtihin four stages of development: the Walking-Horsecar Era (pre-1888), the Electric Streetcar Era (1888-1920), the Recreational Automobile Era (1920-45), and the Freeway Era (1945-present). See Fig. 3-14 (p. 215).

Ghetto (216)

An intraurban region marked by a particular ethnic character. Often an inner-city poverty zone (such as the black ghetto) in the U.S. city, whose residents are involuntarily segregated from other racial and income groups.

Silicon Valley (217)

Popular name of California's Santa Clara Valley near San Francisco, home of the American microprocessor industry and one of the quintessential geographic expressions of postindustrialism on the cultural landscape (see Fig. 3-15, p. 217).

Slow-Growth Epoch (217)

Suggested fifth stage of the Borchert model (see above), spanning the years since 1970, in which higher energy costs and declining birth rates prevail combined with slow overall growth and the increasing dominance of the Sunbelt.

Outer city (218)

The outer metropolitan ring, no longer "sub" to the "urb," which is now the essence of the contemporary U.S. city as it acquires a critical mass of population and activities and becomes co-equal to the central city that spawned it.

Hypothesis of the "galactic metropolis" (218 box)

The notion that today's U.S. urbanization is so far-flung that it is best viewed as a "galaxy" of huge metropolitan concentrations loosely interspersed with more or less continuous clusters of people and buildings.

Suburban downtowns (218)

Huge concentrations of activities in the outer suburban city--usually focused on large regional shopping centers--that are the suburban equivalent of the central-city downtown (CBD).

Urban realms model (219)

Model of contemporary U.S. intraurban spatial structure, dominated by a multiplicity of downtown-level centers (the central-city CBD and a ring of suburban minicities), each serving an independent surrounding sector or realm that is economically self-sufficient. See Fig. 3-17 (p. 219).

Gentrification (220)

The upgrading of older urban residential areas by new higher-income residents, usually displacing from the neighborhood established lower-income inhabitants who increasingly resist this trend toward the disintegration of their communities.

Nativist culture (221)

The complex culture of the United States that matured with the emerging nation, which was best established in suburbia and consists of the following amalgam of values: love of newness, desire to be close to nature, freedom to move, individualism, social acceptance, aggressive pursuit of upward mobility, and a firm sense of destiny.

Dialect (221)

Regional or local variation in the use of a major language, such as the distinctive accents of many residents of the U.S. South or New England.

Toponymy (221)

The study of place names, which provide important clues to the evolution of cultural landscapes.

Mosaic culture (222)

The ongoing new fragmentation of U.S. social groups into smaller and more specialized communities, stratified not only by race and income but also by age, lifestyle, and occupational status.

Mental map (223-224)

The structured spatial information an individual acquires in his or her perception of the surrounding environment and more distant places. A designative "mental map" involves the objective recording of geographical information received (i.e., Florida is a narrow peninsular state at the southeastern corner of the United States). An appraisive "mental map" entails the subjective processing of spatial information according to a person's biases (i.e., Florida is a nice place to live because it has a balmy climate, opportunities for water recreation, and a rapidly growing postindustrial economy).

Vernacular region (224)

Popularity-perceived regions which usually do not coincide with the formal regionalization schemes developed by geographers; examples in the U.S. are "Sunbelt" and "Midwest."

Primary economic activity (226)

Activities engaged in the direct extraction of natural resources from the environment--such as mining, fishing, lumbering, and especially agriculture.

Secondary economic activity (226)

Activities that process raw materials and transform them into finished products; the manufacturing sector.

Tertiary economic activity (226)

Activities that engage in services--such as transportation, banking, retailing, finance, education, and routine office-based jobs.

Quaternary economic activity (226)

Activities engaged in the collection, processing, and manipulation of information.

Quinary economic activity (226)

Managerial or control-function activity associated with decisionmaking in big organizations.

Fossil fuels (227)

Energy resources formed by the geologic compression and transformation of ancient plant and animal organisms--coal, petroleum (oil), and natural gas--which still constitute an overwhelmingly large proportion of the U.S. energy supply.

Economies of scale (231)

The savings that accrue from large-scale production whereby the unit cost of manufacturing decreases as the level of operation enlarges. Supermarkets operate on this principle and are able to charge lower prices than small groceries.

Historical inertia (231)

The need to continue utilizing expensive manufacturing equipment and other capital facilities for their full lifetimes in order to cover initial long-term investments; a major reason for the persistence of the American Manufacturing Belt, despite the obsolescence of much of its physical plant.

Postindustrial economy (232, 233 box)

Emerging economy in the United States and a handful of other highly advanced countries as industry gives way to a high-technology productive complex dominated by services, information-related, and managerial activities.

SELF-TESTING QUESTIONS

Cover this side of the page with a sheet of paper. Uncover each line after you have attempted to answer the question in the left column. If necessary, refer to the textbook page(s) listed at the right.

Question	Answer	Page
URBAN GEOGRAPHY	URBAN GEOGRAPHY	
What is the difference between interurban and intraurban geography?	Interurban geography treats macroscale urban systems, viewing cities as points in an interacting network; intraurban geography treats the internal spatial organization of the individual metropolis.	192
What are the names of the four structural models of intraurban form?	The concentric zone, sector, multiple nuclei, and urban realms models.	192-193, 219
NORTH AMERICAN CHARACTERISTICS	NORTH AMERICAN CHARACTERISTICS	
What is a postindustrial economy?	One dominated by the production and manipulation of information, skilled services, and "high-tech" activities.	191
What are some major geographic differences between the U.S. and Canada?	The U.S. has a much bigger population, but a smaller area; in the U.S. "east" and "west" are the main directions, while in Canada everything is "north" or "south;" the U.S. population is widely dispersed, while the Canadian people are highly concentrated along their southern border.	195
What percentage of the Canadians are of British ancestry? Of French ancestry?	The last census showed that 40 percent were of British ancestry, 27 percent of French origin.	196

Question	Answer	Page
What are Canada's two largest provinces and cities in population size?	Ontario and Quebec; Toronto and Montreal.	197
Where is the heart of Francophone Canada?	Quebec, where over 85 percent of the people speak French.	197
What are the different leading social divisions in the U.S. and Canada?	In Canada, along linguistic and ethnic lines; in the U.S., by race and income.	197
What is meant by the term time-space convergence?	The progressive overcoming of distance since 1800 that allowed far-off places to become ever closer to each other.	198
PHYSICAL GEOGRAPHY	PHYSICAL GEOGRAPHY	
What is a physiographic province?	A physically-uniform region in terms of topography, climate, vegetation, soils, and other environmental variables.	198
What are Arid and Humid America?	The dry and moist halves of the conterminous U.S., roughly divided by the transition zone along the 20-inch (50-centimeter) isohyet.	199
What are the broadest soil and vegetation divisions in the U.S. environment?	Arid America contains alkalinic soils and grassland vegetation; Humid America contains acidic soils and forest vegetation.	200
What is the rain shadow effect?	The existence of dry environmental conditions downwind from mountain ranges as a result of orographic precipitation.	201-1 box
Name the five Great Lakes and their outlet to the sea.	Lakes Superior, Michigan, Huron, Erie, Ontario; the St. Lawrence River.	201, 202 map
Name the major tributaries of the Mississippi Rivier.	The Missouri, Ohio, Tennessee, and Arkansas Rivers.	201
What is an aquifer?	An underground, porous, water-bearing rock layer, tapped by surface wells.	203
What is meant by the term smog?	A contraction of the words "smoke" and "fog," a severe form of air pollution caused by industrial and vehicle emissions, intensified greatly in places where stagnant air is common.	203

POPULATION AND URBANIZATION	POPULATION AND URBANIZATION	
How often, on the average, do Americans relocate to a new residence?	Eighteen percent of the U.S. population relocates each year, or, put differently, people move about once every five years.	206 box
How does U.S. population mobility compare to the rest of the world?	The U.S. is way ahead: whereas Americans move 14 times during an average lifetime, the British relocate only eight times, the French seven times, and the Japanese five times.	207 box
What were the three separate <u>culture hearths</u> of the United States?	New England, the Middle Atlantic area, and Tidewater Maryland and Virginia.	207-8
What was the "Great American Desert"?	The steppelands west of 100°W longitude which were first thought to be drier than they really are, causing the frontier to leapfrog to the Pacific coast and only fill in the bypassed areas later.	208
What was the Township-and-Range land division system?	A surveying scheme that facilitated the easy separation of properties along straight lines resulting in a checkerboardlike cultural landscape.	209
When did the Industrial Revolution occur in the U.S.?	Between the Civil War and 1920.	209
Name the four stages in Borchert's model of American metropolitan evolution.	The Sail-Wagon Epoch (1790-1830), the Iron Horse Epoch (1830-70), the Steel-Rail Epoch (1870-1920), and the Auto-Air-Amenity Epoch (1920-70).	210-11
What is the <u>core area</u> of the realm, and what are the four corners of the rectangular boundary that encloses this region?	The American Manufacturing Belt; St. Louis, Milwaukee, Boston, and Washington, D.C.	211-14
What is <u>central place theory</u>?	A location theory explaining the spacing of urban centers of differing size and the forces that govern their distributions.	212-13 box
What is the <u>range</u> of a good?	The maximum distance that people are willing to travel to purchase a central good or service.	213 box
What is an urban hierarchy?	The rank ordering of central places in a national urban system with the highest-order center at the top, followed in descending order by smaller cities, towns, villages, and hamlets.	213 box

What is the economic heartland of this core region?	The Boston-Washington Megalopolis (its Canadian equivalent is Main Street, stretching from Quebec City southwest to Windsor).	214
Name the four stages in the Adams model of intraurban spatial evolution.	The Walking-Horsecar Era (pre-1888), the Electric Streetcar Era (1888-1920), the Recreational Auto Era (1920-1945), and the Freeway Era (1945-present).	215-16
What is meant by the "Slow-Growth" Epoch?	The proposed fifth stage of the Borchert model (post-1970), marked by slower population growth, much higher energy prices, the renewed growth of rural America, and the continued rapid rise of the Sunbelt.	217
What is meant by the terms outer city, suburban downtowns, and the urban realms model?	The newly urbanized suburban ring, now coequal to the central city that spawned it; suburban downtowns anchor the outer ring of the now multi-centered metropolis; Vance's generalization that views today's metropolis as consisting of several self-sufficient sectors, each organized around its own downtown.	218-19
What is the "hypothesis of the galactic metropolis"?	That today's U.S. urbanization comprises one vast galaxy of huge metropolitan concentrations, loosely interconnected by separated smaller clusters of people and buildings.	218 box
CULTURAL GEOGRAPHY	CULTURAL GEOGRAPHY	
What are the values that underlie the American nativist culture?	Love of newness, desire to be near nature, freedom to move, individualism, societal acceptance, aggressive pursuit of upward mobility, and a firm sense of destiny.	221
What is a dialect?	A local variation in a language spoken across a large area.	221
What is the emerging mosaic culture?	The new ongoing fragmentation of American society into a plethora of narrowly-defined communities, not only along income, racial, and ethnic lines, but also according to age and lifestyle.	222-23
What is the difference between a designative and an appraisive mental map?	Designative mental maps involve the objective placement of spatial information on a map; an appraisive mental map involves the manipulation of information according to one's personal biases.	223-25

ECONOMIC GEOGRAPHY	ECONOMIC GEOGRAPHY	
Identify primary, secondary, tertiary, quaternary, and quinary economic activity.	The extractive sector, especially mining and agriculture; manufacturing; the services sector; the information sector; managerial decision making in big organizations.	226
Name the three fossil-fuels and their three leading North American source areas.	Coal--Appalachia, the northern Great Plains, Midcontinent; petroleum--Gulf Coast, Midcontinent, and Alaska; natural gas--Gulf Coast, Midcontinent, and Appalachia.	227
Describe the application of the Von Thunen Model to the explanation of the spatial structure of U.S. agriculture.	Like modern Europe, the model fits at the macroscale, with agricultural regions of decreasing intensity concentrically arranged with increasing distance from the central "supercity"--Megalopolis.	229-30
What are economies of scale?	In industrial location, the savings that accrue from large-scale production whereby the unit cost of manufacturing decreases as the level of the operation enlarges.	231
How does the principle of historical inertia help to maintain the aging industries of the Manufacturing Belt?	Because of the need to continue using expensive manufacturing facilities for their full lifetimes to cover long-term investments that were made decades ago.	231
What are the location factors that attract today's pacesetting "high-tech" companies?	A university, urban life, highly skilled labor, sunshine, recreational water, good housing, luxury housing for executives, investment capital, risk reduction, and local cooperation.	232-33
Which U.S. states were the biggest gainers and losers in per-capita income change from 1975 to 1982?	As shown in Fig. 3-27 (p. 234), the winners were Texas, Louisiana, Oklahoma, Connecticut, Florida, Colorado, California, Wyoming, Virginia, and upper New England; the losers were heavily clustered within the American Manufacturing Belt.	233-34
NORTH AMERICAN REGIONS	NORTH AMERICAN REGIONS	
What is the Anglo-American Core?	The Manufacturing Belt, now facing problems in keeping abreast of the realmwide transition from industrial to postindustrial power.	235-36

Question	Answer	Page
What problems do northern New England and Maritime Canada share?	Both are generally rural, possess difficult environments, and were historically bypassed in favor of more dynamic and fertile inland areas; although agriculture and fishing are not growth industries any longer, tourism is and presents an opportunity in this scenic region.	236-37
What was the outcome of 1970s separatist movement in Quebec?	The French Canadians gained most of their goals by eliminating English domination of Quebec's culture; the 1980 referendum soundly defeated the idea of secession from Canada.	237-38
What are the South's persistent economic problems?	Uneven development has favored certain areas, left others untouched by recent progress; much growth has occurred at the edges of the region, and Southern central cities increasingly face the problems of their Northern counterparts while their suburbs thrive.	240
What are the tricultural influences evident in the Southwest?	The growing Anglo influence, the persistently Hispanic flavor of local cultures, and the sporadic native American (Indian) presence.	241
What are the two leading resources of the Interior Periphery?	Its rich mineral supplies and its open-space amenities--with both attracting population growth.	241-42
Name the "Nine Nations" of North America and their approximate geographic extents.	The Foundry (approximately the Manufacturing Belt), New England (all 6 states plus the Maritimes), Quebec (the entire province), Dixie (concurs with the South mostly), the Islands (southern Florida and the Caribbean), the Breadbasket (analogous to the Agricultural Heartland), the Empty Quarter (similar to Interior Periphery), Ecotopia (generally, the Pacific Northwest), and Mex-America (the broad border zone from California to Texas).	244

MAP EXERCISES

MAP COMPARISON

1. The Canadian population has always clung closely to the country's southern border. By comparing the maps of North America (Fig. 3-4, p. 196), 1980 Population Distribution (Fig. 3-9, p. 205), and Expansion of Settlement (Fig. 3-10, p. 208), record your observations in detail about this spatial pattern over the past two centuries.

2. Compare the map of North American physiography (Fig. 3-5, p. 202) and settlement expansion (Fig. 3-10, p. 208). What relationships are apparent in attraction and avoidance between landform regions and population patterns? Differentiate between 19th and 20th century settlement patterns, and single out those areas where a reversal of settlement preferences seem to have occurred after 1900.

3. Compare the map of geographic patterns of cancer (Fig. 3-7, p. 204) with all the other maps in this chapter. What correspondences, if any, can be observed? As an extra exercise, review the industrial maps in either the National Atlas of the United States of America or the Oxford Regional Economic Atlas of the United States and Canada (cited on p. 247; both should be available in your library's reference department) and offer further observations about the possible relationships between certain cancers and manufacturing operations.

4. Carefully examine the map of the Continental Core Region (Fig. 3-12, p. 211), and offer observations about the interregional flows of railroad freight traffic in North America (line symbols on maps was a topic introduced earlier on p. 53 of the text). This map is of mid-1970s vintage: given the recent changes in the geography of manufacturing and goods movement in the U.S., what changes would you expect to see on a late 1980s map of rail freight flows?

5. Study the map of North America's leading deposits of fossil fuels (Fig. 3-23, p. 228). Learn the distributions of each energy resource by listing the five major regions where coal, oil, and natural gas are each produced.

6. Examine the map of the future residential preferences of University of Miami students (Fig. 3-22, p. 226). Do you and your classmates agree or disagree with these choices? How might these choices differ if the map information was collected at a university in California, Alabama, New Jersey, or Illinois? How might these choices change over time: is your own list of most preferred states the same today as it was in 1980?

7. Compare the map of the text regionalization scheme (Fig. 3-28, p. 235) with the "nine-nations" map (Fig. 3-34, p. 245). What differences and similarities are observed?

MAP CONSTRUCTION

(Use outline maps at the end of this chapter)

1. In order to become more familiar with the North American environment, place the following physical-geographic information on the first outline map:

   a. Rivers: Connecticut, Hudson, Delaware, Susquehanna, Potomac, Ohio, Tennessee, Mississippi, Missouri, Arkansas, Platte, Red River of the North, Rio Grande, Colorado, Snake, Columbia, Willamette, Sacramento, San Joaquin, St. Lawrence, Fraser, Okanagan

   b. Water bodies: Chesapeake Bay, Gulf of Mexico, San Francisco Bay, Puget Sound, Great Salt Lake, Lake Superior, Lake Michigan, Lake Huron, Lake Erie, Lake Ontario, Bay of Fundy, Gulf of St. Lawrence, Lake Winnipeg, Strait of Juan de Fuca

c. Land bodies: Cape Cod, Long Island, Delmarva Peninsula, Cape Hatteras, Florida Keys, Mississippi Delta, Mohave Desert, Olympic Peninsula, Vancouver Island, Grand Canyon, Alaskan Peninsula, Aleutian Islands, Newfoundland

d. Mountains: Laurentians, Appalachians, Green Mountains, White Mountains, Adirondacks, Great Smoky Mountains, Ozark Plateau, Rocky Mountains, Wasatch Mountains, Sierra Nevada, Cascades, Klamath Mountains, Olympic Mountains, Pacific Coast Ranges, Black Hills, Tehachapi Mountains.

2. On the second map, political-cultural information should be entered as follows:

   a. Write in the name of each U.S. state and Canadian province

   b. Draw in the major concentrations of minority populations, using Fig. 3-20 (text p. 223) as a guide

3. On the third outline map, urban-economic information should be entered as follows:

   a. Cities: (locate and label with the symbol •): Boston, Hartford, New York, Buffalo, Philadelphia, Pittsburgh, Baltimore, Washington, D.C., Richmond, Norfolk, Raleigh, Charlotte, Atlanta, Charleston (S.C.), Savannah, Jacksonville, Tampa, Orlando, Miami, Mobile, Birmingham, New Orleans, Memphis, Nashville, Louisville, Cincinnati, Cleveland, Indianapolis, Detroit, Milwaukee, Chicago, Minneapolis, St. Louis, Des Moines, Kansas City, Omaha, Oklahoma City, Tulsa, Houston, Dallas, San Antonio, Austin, El Paso, Albuquerque, Denver, Salt Lake City, Tucson, Phoenix, Las Vegas, San Diego, Los Angeles, San Francisco, Portland (Ore.), Seattle, Vancouver, Calgary, Regina, Winnipeg, Windsor, Toronto, Ottawa, Montreal, Quebec City, Halifax

   b. Economic regions (identify with circled letter):

      A - Silicon Valley
      B - Atlantic Seaboard Megalopolis
      C - Corn Belt
      D - Main Street conurbation
      E - Mesabi Iron Range
      F - Powder River Basin
      G - Alaskan North Slope
      H - Research Triangle
      I - Boundary between Arid and Humid America (draw in)

PRACTICE EXAMINATION

SHORT-ANSWER QUESTIONS

Multiple Choice

1. Which of the following is not a "corner" city of the American Manufacturing Belt (Continental Core Region):

    a) St. Louis
    b) Boston
    c) Milwaukee
    d) Philadelphia
    e) Washington, D.C.

2. Which city is located closest to the Canadian capital of Ottawa:

    a) Toronto
    b) Windsor
    c) Vancouver
    d) Minneapolis
    e) Washington, D.C.

3. Which of the following is a secondary economic activity:

    a) iron mining
    b) beer brewing
    c) retail sales
    d) managing a corporation
    e) cotton farming

True-False

1. Manitoba is one of Canada's "prairie" provinces.

2. Gentrification is a term applied to the resuscitation of an aging "smokestack" industry.

3. The shaping of the spatial form of the American metropolis during the 1920s was more heavily influenced by private automobiles than public transit systems.

Fill Ins

1. The Atlantic Seaboard Megalopolis stretches northeastward from Washington, D.C. to the vicinity of the metropolitan area of ________________.

2. The northward extension of the Sierra Nevada through the states of Oregon and Washington is called the ________________ Mountains.

3. The outlet to the sea for the Great Lakes is the ________________ River.

Matching Question on States and Provinces

_____	1.	Maritime province	A.	Washington
_____	2.	Corn Belt	B.	North Carolina
_____	3.	Contains "gateway" to Latin America	C.	California
_____	4.	Oil shale	D.	New York
_____	5.	North Slope oilfields	E.	British Columbia
_____	6.	Research Triangle	F.	Florida
_____	7.	Adirondacks	G.	Nova Scotia
_____	8.	San Andreas fault	H.	Quebec
_____	9.	Francophone Canada	I.	Illinois
_____	10.	Mt. St. Helens	J.	Ontario
_____	11.	Site of Canadian capital	K.	Texas
_____	12.	Winter Wheat Belt	L.	Kansas
_____	13.	Vancouver	M.	Colorado
_____	14.	"Silicon Prairie"	N.	Louisiana
_____	15.	Mississippi Delta	O.	Alaska

ESSAY QUESTIONS

1. The point was made that one of North America's greatest achievements was to overcome the "tyranny" of distance--a spatial quality the Soviets must view with the greatest of envy. Discuss how the United States and Canada overcame their physical barriers to achieve the significant time-space convergence that allowed their transcontinental economies and societies to emerge. Compare and contrast this experience in effective long-distance spatial organization with that of the Soviet Union (referring back to the preceding chapter if necessary).

2. Draw a sketch map of the agricultural regions of the United States, and explain its spatial structuring within the framework of the macro-Thunian model.

3. Imagine that you are the chief executive of a successful young company that manufactures software for the newest high-speed computers. Where would you locate your company and why? Establish a "short list" of three possible sites, and choose one by a process of elimination in which your decision is built upon the most important locational variables for your plant and its highly skilled workforce.

4. Most of the western half of the United States is part of Arid America. Explain why this dry environment exists, and what its consequences are for climate, vegetation, and soils. Also, discuss the implications of this situation for the human use of the earth, and speculate what kind of economy might have arisen had the country been settled from the Pacific coast moving inland to the east.

5. The American central city has undergone a steady transformation over the past two centuries. Discuss its spatial evolution through the four stages of the Adams model, emphasizing the concentric-zone, sector, and multiple nuclei expressions in the first half of this century. Then, briefly review the declining role of the big city in the age of the postindustrial multinodal metropolis, stressing the context of the urban realms model and the so-called "gentrification" movement.

TERM PAPER POINTERS

The "Term Paper Pointers" section of Chapter 1 in this Study Guide offered suggestions about approaching research and writing on geographic realms and their components, and you may wish to consult that material if you are undertaking a report on a North American region.

Once again, the complexity of North America makes a good regional geography an appropriate place to begin your reading. Among the current titles recommended from the citations on text pp. 246-247 are: Birdsall & Florin, Hart [1972], McCann (a useful newer book on Canada only), Paterson, Putnam & Putnam (Canada only), and White, Foscue & McKnight. A goodly number of atlas-type publications are also worth consulting for certain topics, with many containing substantial texts in addition to map collections: Allen & Turner, Atlas of North America, Atlas of the U.S., Cuff & Young, "Energy:..," National Atlas of the U.S.A., Oxford Regional Economic Atlas . . ., and Rooney et al. Several helpful topical treatments can be found as well. For physical geography see Atwood, Hunt, and Watson & O'Riordan; cultural and social geography are covered in Bell, Berry, Brunn, Gastil, Herbers, "Immigrants," Jackson, Knox, Kowinski, Louv, Malcolm, McKee, Rooney et al., Ross & Moore, Thernstrom, Watson, and Zelinsky [1973]; for perceptual and landscape geography see Clay, Gould & White, Lewis [1983c] Meinig [1979], Parfit, Stewart, Vale & Vale, and Zelinsky [1980]; economic geography is treated in Clark, Cutter, Hall & Markusen, Hart [1975], Leinberger & Lockwood, Malecki, Markusen, Schertz et al., and Wheeler & Muller; and historical geography is covered in Meinig [1986], Mitchell & Groves, Vance, both entries by Ward, and by one of the great classic works of scholarship in this century--Brown--still available in most libraries. Regional studies of subnational areas are treated in Boles, Borchert [1987], Browning, Raitz et al., Reisner, Saxenian, Weinstein, and Yeates [1975].

The urban geography Systematic Essay is also a good springboard for a research paper because a vast literature exists, much of it accessible to newcomers. Among the many first-rate current texts are Brunn & Wheeler, Christian & Harper, Hartshorn (second revised edition due in 1989--1980 edition available until then), Knox, and Palm; the reader edited by Bourne is also a fine background source. Among the more specialized works in historical urban geography are Borchert [1967], Chudacoff, Gottmann, Jackson, Mayer, Ward [1971]--and the monumental volume by Vance. Contemporary topics (and city regions) are covered in Adams [1976], Baerwald, Bernard & Rice, Goldberg & Mercer, King, Lewis [1983 a & b], Muller [1981], Noyelle & Stanback, Phillips & Brunn, and Yeates [1980].

SPECIAL EXERCISE

Model Box 3 (pp. 212-213) treats central place theory, one of geography's most important location theories, which helps us to understand why urban centers of varying size are distributed the way they are. This special exercise requires your finding and reading a more elaborate introduction to the subject, which provides a deeper understanding of its basic concepts. The article in question should be readily available in your library: Arthur & Judith Getis, "Christaller's Central Place Theory," Journal of Geography, 65 (May 1966), 220-226.

Reread Model Box 3 and the Getis & Getis article, and then answer the following questions.

a. Are all urban settlements central places? What kinds of urban centers are not? List ten urban settlements in your home area, and classify them according to central-place or dispersed-place status.

b. What is the difference between the upper and lower limits in the range of a good? How do the terms "ideal limit" and "real limit" differ? Use the diagram on p. 221 of the Journal to support your answers.

c. What is a complementary region? Under what conditions does it assume a circular shape? How do hexagonal patterns develop?

d. Briefly list the ideal conditions about the landscape on which a central-place system develops. What behavioral constraints exist concerning the shopping desires of residents of this region?

e. Describe in your own words how the k=3 central-place network is derived, supporting your answer by referring to the diagrams on Journal pp. 223-225.

f. Empirical (real-world) applications of central place theory are often undertaken (the text raises this issue again in Chapter 9). For the United States, several expressions can be found. If you can, obtain a Road Atlas and find the map of Iowa. This Corn Belt state meets many of the ideal conditions enunciated by Christaller: carefully observe the spacing of cities, towns, and villages, and record your conclusion as to whether a k=3 network is apparent. Now turn to Kentucky and see what evidence can be found about the actual "packing together" of local regions: with the exception of counties touching that state's boundary, count for each county the number of bordering counties (the average number in a Christaller system should be 6.0 contacts--average Kentucky's counties and see what you get). This procedure can be extended to other states where the Township-and-Range survey system did not bias the number of county contacts; some good ones to try are North Carolina, Virginia, Pennsylvania, New York, New Jersey, and the New England states.

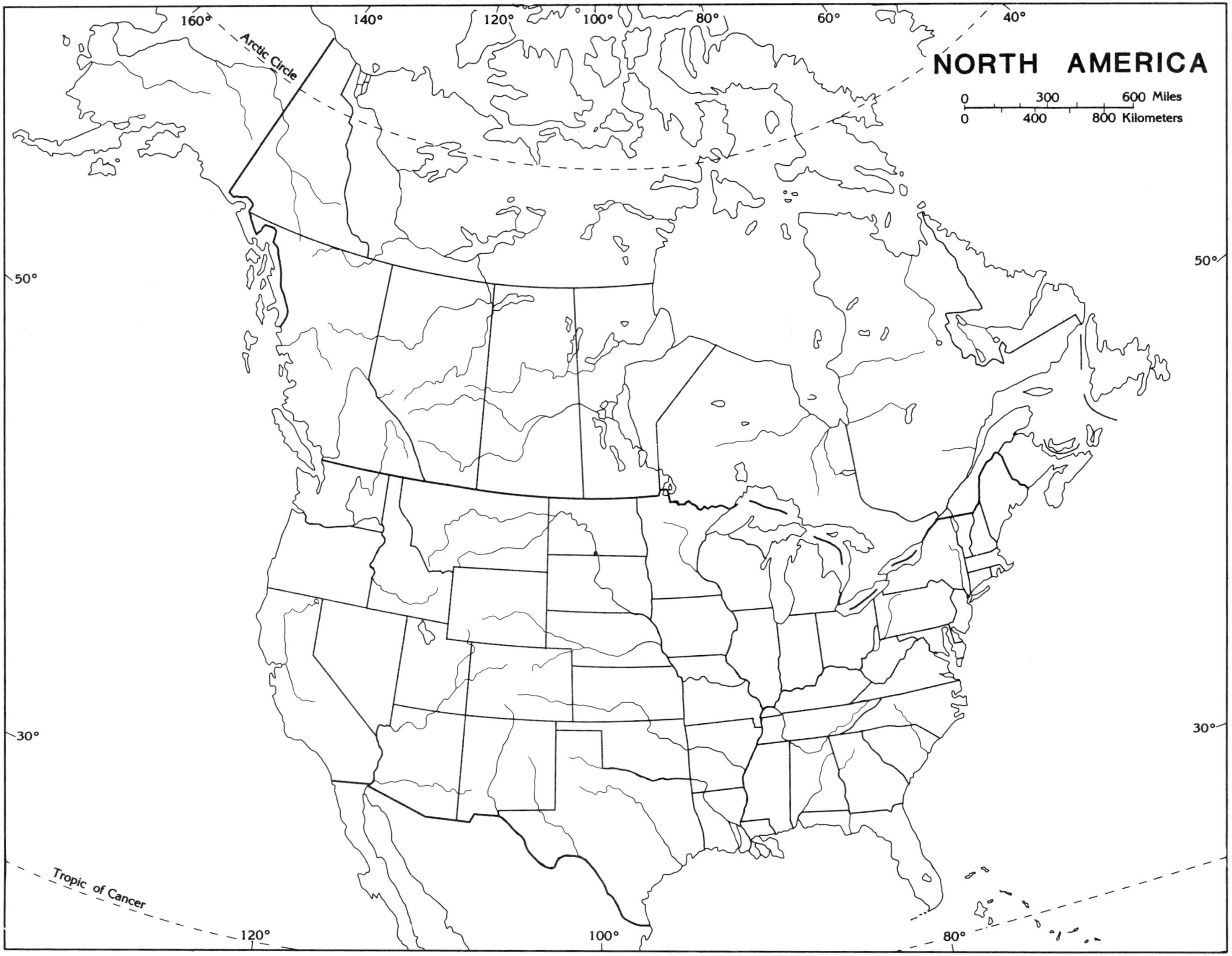

NORTH AMERICA
0 300 600 Miles
0 400 800 Kilometers
160°
140°
120°
100°
80°
60°
40°
50°
30°
Arctic Circle
Tropic of Cancer
120°
100°
80°

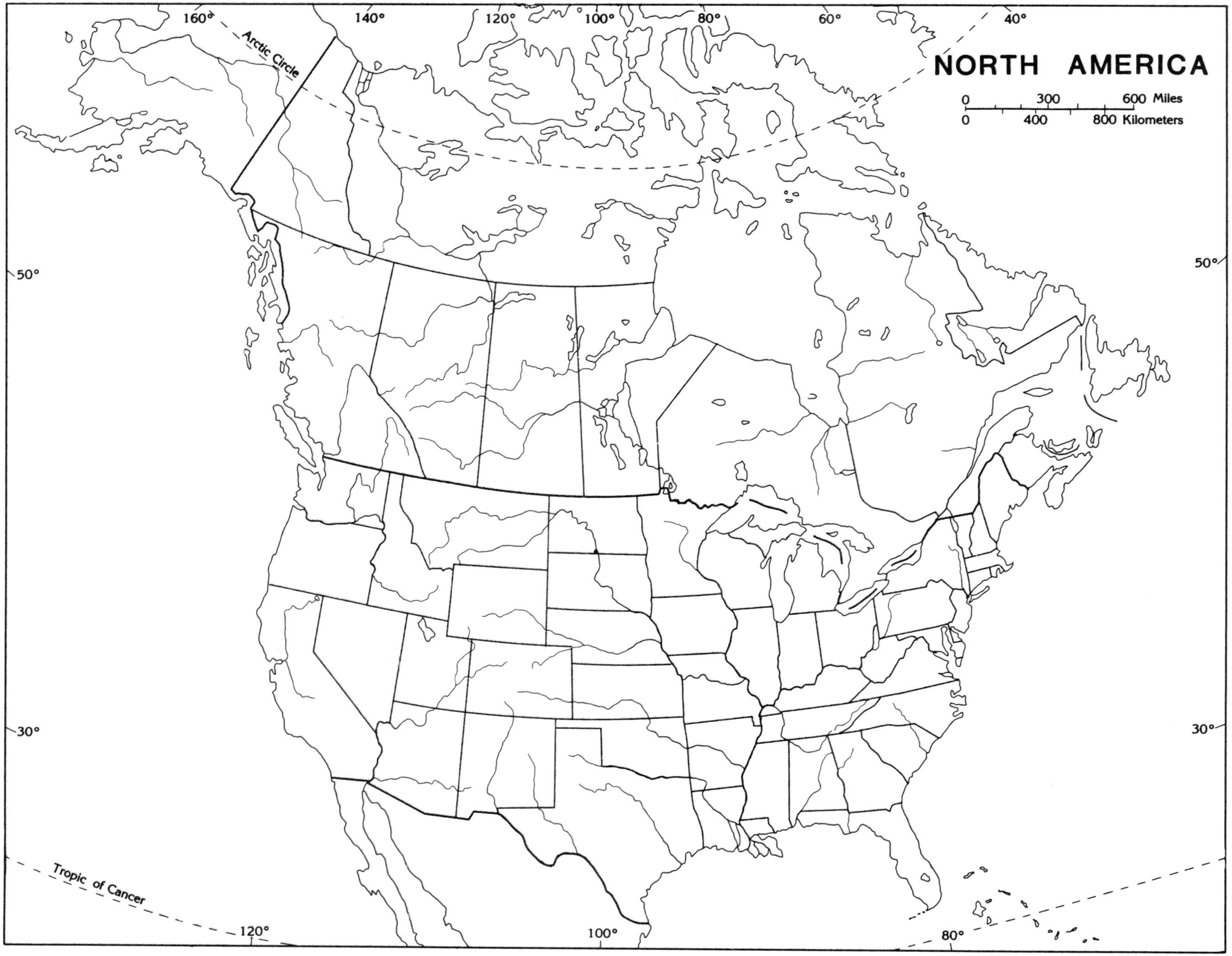
NORTH AMERICA
0 300 600 Miles
0 400 800 Kilometers
160°
140°
120°
100°
80°
60°
40°
Arctic Circle
50°
50°
30°
30°
Tropic of Cancer
120°
100°
80°

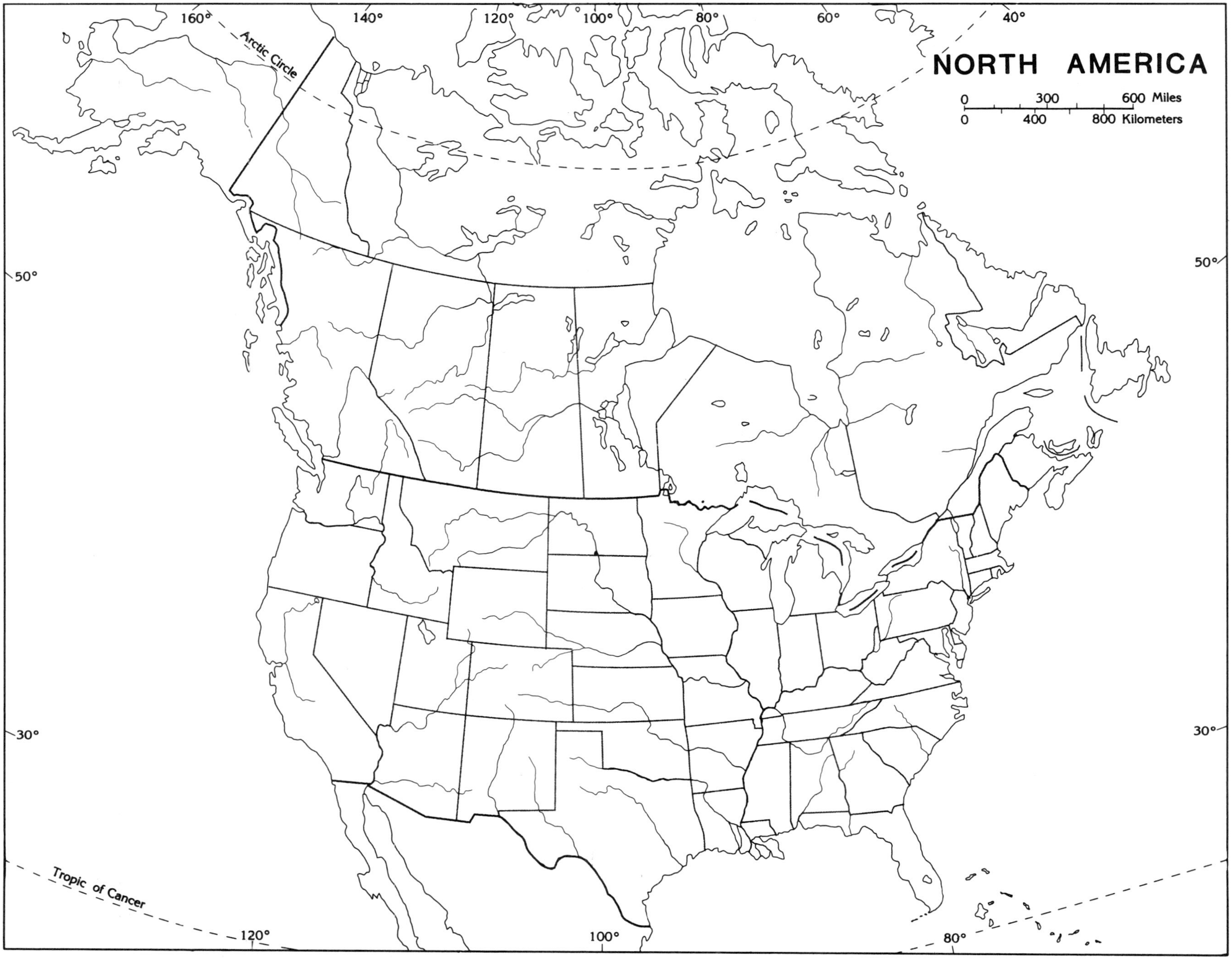
NORTH AMERICA
0 300 600 Miles
0 400 800 Kilometers
Arctic Circle
Tropic of Cancer
160°
140°
120°
100°
80°
60°
40°
50°
30°
50°
30°
120°
100°
80°

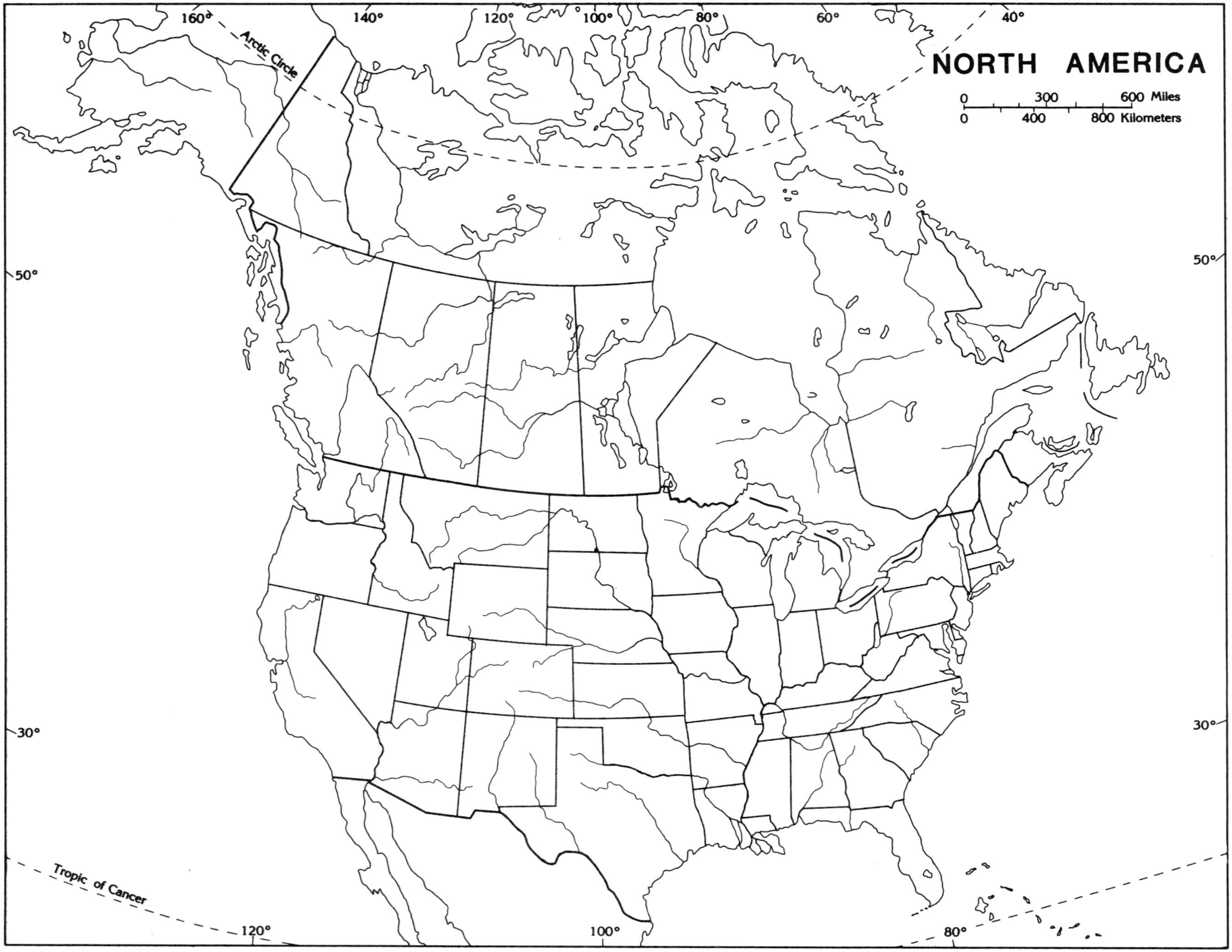
NORTH AMERICA
0 300 600 Miles
0 400 800 Kilometers
160°
140°
120°
100°
80°
60°
40°
50°
30°
Arctic Circle
Tropic of Cancer

VIGNETTE: PRODIGIOUS JAPAN--TRIUMPH OF TECHNOLOGY

OBJECTIVES OF THIS VIGNETTE

This Vignette treats Japan, which is rightfully classified with the developed realms and especially with North America--whose budding postindustrial qualities it shares. After a brief introductory overview, Japan's modern history and economic development are reviewed against a domestic setting of rather limited environmental resources. The country's remarkable industrialization and modernization trends are treated next, emphasizing economic spatial organization. Food production systems are then covered, and a concluding section looks at the Japanese leadership role--unintended though it was--in the postindustrial age of the late twentieth century.

Having learned the regional geography of Japan, you should be able to:

1. Understand its place in today's world.

2. Describe the country's stormy political history since 1900.

3. Trace the modernization trend and its impact from the Meiji Restoration to the present.

4. Describe the country's resource base and the broad regional structuring of its productive activities.

5. Understand the Japanese food production system and its geographic foundations.

6. Grasp the cultural and social dilemmas Japan now faces as the world looks to it for a greater measure of technological and innovation leadership.

7. Locate the leading physical, cultural, and economic-spatial features of this realm on an outline map.

GLOSSARY

Archipelago (248)

A chain of islands grouped closely together--as in the case of Japan.

Areal functional organization (255)

Allen Philbrick's model of regional organization based on 5 interrelated ideas that activities concentrate in a locality, interact with other places, and participate in the evolution of a multi-level hierarchy of areal organization.

Physiological density (259)

The number of persons per unit of arable or cultivable land; Japan exhibits one of the world's highest figures more than 5 times higher than India's) exacerbated by the large amount of Japanese territory that cannot be utilized for agriculture.

Aquaculture (262)

The use of a pond or other artificial body of water to grow food products, including fish, shellfish, and even seaweed; method often also used in shallow bays and estuaries.

SELF-TESTING QUESTIONS

Cover this side of the page with a sheet of paper. Uncover each line after you have attempted to answer the question in the left column. If necessary, refer to textbook page(s) listed at the right.

Question	Answer	Page
JAPAN'S CHARACTERISTICS	JAPAN'S CHARACTERISTICS	
Name the four main islands of the Japanese archipelago.	Honshu (the largest), Kyushu, Shikoku, and Hokkaido.	248
How does Japan's population help it qualify as a separate realm?	Its size (ca. 123 million), and particularly its remarkable homogeneity.	249
Which nation forced open the modern era of Japanese international contact and trade?	The United States, in 1853; Commodore Perry's flotilla, through a show of naval strength, extracted the necessary agreements.	251
MODERNIZATION	MODERNIZATION	
When were the old rulers of Japan overthrown, paving the way for reform?	In 1868, during the rebellion called the Meiji Restoration--the return of enlightened rule.	251
How were the Japanese able to transform their society since 1868?	By careful domestic planning rather than through foreign intrusion; building on internal resources, the Japanese expanded to create an empire.	251-53

Question	Answer	Page
What are Japan's shortcomings in local industrial resources?	Inferior coal, and very little iron ore; oil supplies are practically nonexistent, forcing importation and heavy reliance on hydroelectric and nuclear energy production.	253
SPATIAL ORGANIZATION	SPATIAL ORGANIZATION	
Where did Japan's early industrial growth concentrate and why?	Largely in the coastal cities, because raw materials were transported by water and labor forces were already located there.	255
List the main ideas of Allen Philbrick's <u>areal functional organization</u>.	Human activities concentrate in a locality, interact with other places, and participate in the evolution of a multi-level hierarchy of areal organization.	255
What are the leading features of the Kanto Plain?	Japan's dominant urban and industrial region; focused on Tokyo-Yokohama; fine harbor, good climate for farming, and central location in country.	256
What are the leading features of the Kinki District?	Japan's second-ranking economic region; focused on the Kobe-Osaka-Kyoto triangle; good farming area, especially for rice.	257
What are the leading features of the Nobi Plain?	Third-ranking producing region; focused on Nagoya, leading textile manufacturer; may be coalescing with Kobe-Osaka.	257
What are the leading features of the Kitakyushu conurbation?	Japan's fourth largest economic region; focused on 5 cities at northern end of Kyushu Island; steel is major industry; excellently situated on trade routes to Korean and Chinese ports.	257-58
FOOD PRODUCTION	FOOD PRODUCTION	
How has the Japanese agricultural labor force changed since 1920?	As the economy took off, agricultural employment fell from 51 percent of the national total in 1920 to about 9 percent in the 1980s--with nearly nine-tenths today only part-time farmers.	259
Why must Japan rely so heavily on food imports?	Like Britain, arable land is in short supply; despite high-technology, one-third of Japan's food is imported each year--still a remarkably low proportion given such meager domestic agricultural resources.	259-61

What new food resources are being developed?	Besides the unending search for more efficient farming methods, aquaculture is increasingly used to supplement the food supply.	261-62
POSTINDUSTRIAL JAPAN	POSTINDUSTRIAL JAPAN	
What signs of postindustrialism mark Japan in the 1980s?	The rise of an information-based economy--with about 60 percent of the labor force now employed in the tertiary-quaternary-quinary sectors; ultramodern communications networks; computerization of workplaces; the growth of research-based new cities such as Tsukuba Science City.	262
How well-equipped is Japanese culture to handle a world leadership position in technological and other innovations?	There are many doubts, because the Japanese possess something of a national inferiority complex; they much prefer to mimic ideas rather than innovate them (as the photo on text p. 248 so strikingly demonstrates), but continued future success may ultimately depend on seizing a leadership role.	263-64

MAP EXERCISES

MAP COMPARISON

1. The mountainous character of the Japanese archipelago is a dominant feature of the Land and Livelihoods map (Fig. J-2, p. 250). Compare this map to the series of world physical maps (Figs. I-9 to I-13, pp. 14-25) and record your observations about the impact of the highlands on climate, precipitation, vegetation, and soils.

2. Compare the map of Japanese manufacturing regions (Fig. J-4, p. 256) with the map of Land and Livelihoods (Fig. J-2, p. 250). Record your observations as to the correspondence between population distribution and the location of the country's primary, secondary, and tertiary industrial regions.

MAP CONSTRUCTION

(Use outline maps at the end of this vignette)

1. To familiarize yourself with Japan's physical geography, enter the following on the first of the outline maps:

   a. Water bodies: Sea of Japan, Seto Inland Sea (Setonaikai), Korea Strait, Strait of Shimonoseki, Tokyo Bay

b. Land areas: Honshu, Kyushu, Shikoku, Hokkaido, Kinki District, Kanto Plain, Nobi Plain, Sen-en Coastal Strip

2. On the second outline map, enter the following urban-economic information:

a. Cities (locate and label with the symbol •): Tokyo, Yokohama, Osaka, Kobe, Kyoto, Nagoya, Kitakyushu, Fukuoka, Hiroshima, Nagasaki, Sapporo

b. Economic centers (identify with circled letter):

A - Kanto Plain
B - Kinki District
C - Nobi Plain
D - Kitakyushu

PRACTICE EXAMINATION

SHORT-ANSWER QUESTIONS

Multiple Choice

1. Which of the following is located on the island of Kyushu:

a) Yokohama
b) Kitakyushu
c) Nobi Plain
d) Kyoto
e) Honshu

2. Before the Meiji Restoration, Japan's capital city was:

a) Kyoto
b) Tokyo
c) Edo
d) Kinki
e) Honshu

True-False

1. The island of Hokkaido lies outside of Japan's core area.

2. Despite industrialization, Japan still has a large agricultural sector that employs more than one-third of the labor force.

Fill Ins

1. Japan's dominant economic region, which contains its largest city, is the _______ Plain.

2. The number of people per total arable land area is known as a country's _______ density.

Matching Question on the Japanese Realm

_____	1.	Sakhalin Island	A.	Aquaculture
_____	2.	National belief system	B.	Edo
_____	3.	Target of nuclear bomb	C.	Sen-en Coastal Strip
_____	4.	Soviet hegemony	D.	Karafuto
_____	5.	Kyushu conurbation	E.	Tsukuba
_____	6.	Tokaido megalopolis	F.	Shinto
_____	7.	Fish farming	G.	Kurile Islands
_____	8.	Leading textile producer	H.	Nagoya
_____	9.	Japan's Research Triangle Park	I.	Nagasaki
_____	10.	Former name of Tokyo	J.	Kitakyushu

ESSAY QUESTIONS

1. Trace the rise of Japanese empire-building and imperialism from the Meiji Restoration through World War II, and discuss why it was relatively easy for Japan to acquire such vast territorial holdings during the first half of this century.

2. Japan's record of modernization and industrialization over the past century is certainly one of the world's greatest national success stories. Trace the emergence of Japan as an industrial power, emphasizing resources, economic spatial organization, and international trade opportunities; also, comment upon the Japanese crusade to build a pre-World War II empire and the ways in which such imperialism affected domestic industrial growth.

TERM PAPER POINTERS

The literature cited on p. 264-265 provides a good introduction to the realm, with the Time issue of August 1, 1983 ("Japan: A Nation in Search of Itself") offering an especially wide-ranging background on most aspects of Japanese life. Though dated, the 1979 bibliography by Kornhauser is still a valuable guide to the geographical works on Japan. Japanese history is covered in Beasley, Hane, and Tiedemann. For cultural topics, the books by Christopher and Higuchi should prove insightful; Morton, Reischauer, and Scheiner offer additional interpretations. Any research on the economic geography of Japan should begin with the splendid survey by Harris. Industrialization is covered in Murata & Ota, Patrick, and Smith; if your library has a copy, the now-out-of-print book by Robert Hall offers a fine account of Japan in the postwar period. Urban geography is well represented in Kornhauser [1982], with good case studies offered by Allinson [both sources], Eyre, P. Hall, Ito & Nagashima, and Popham. The new postindustrial age is reviewed in Burks, Pollack, Takahashi, and Tatsuno. The most up-to-date general geographies are Japan and MacDonald; useful older titles are Association of Japanese Geographers, Dempster, Pezeu-Massabuau, Pitts, and Trewartha. Two fine atlases are also included in the listing.

There is no Special Exercise for this Vignette

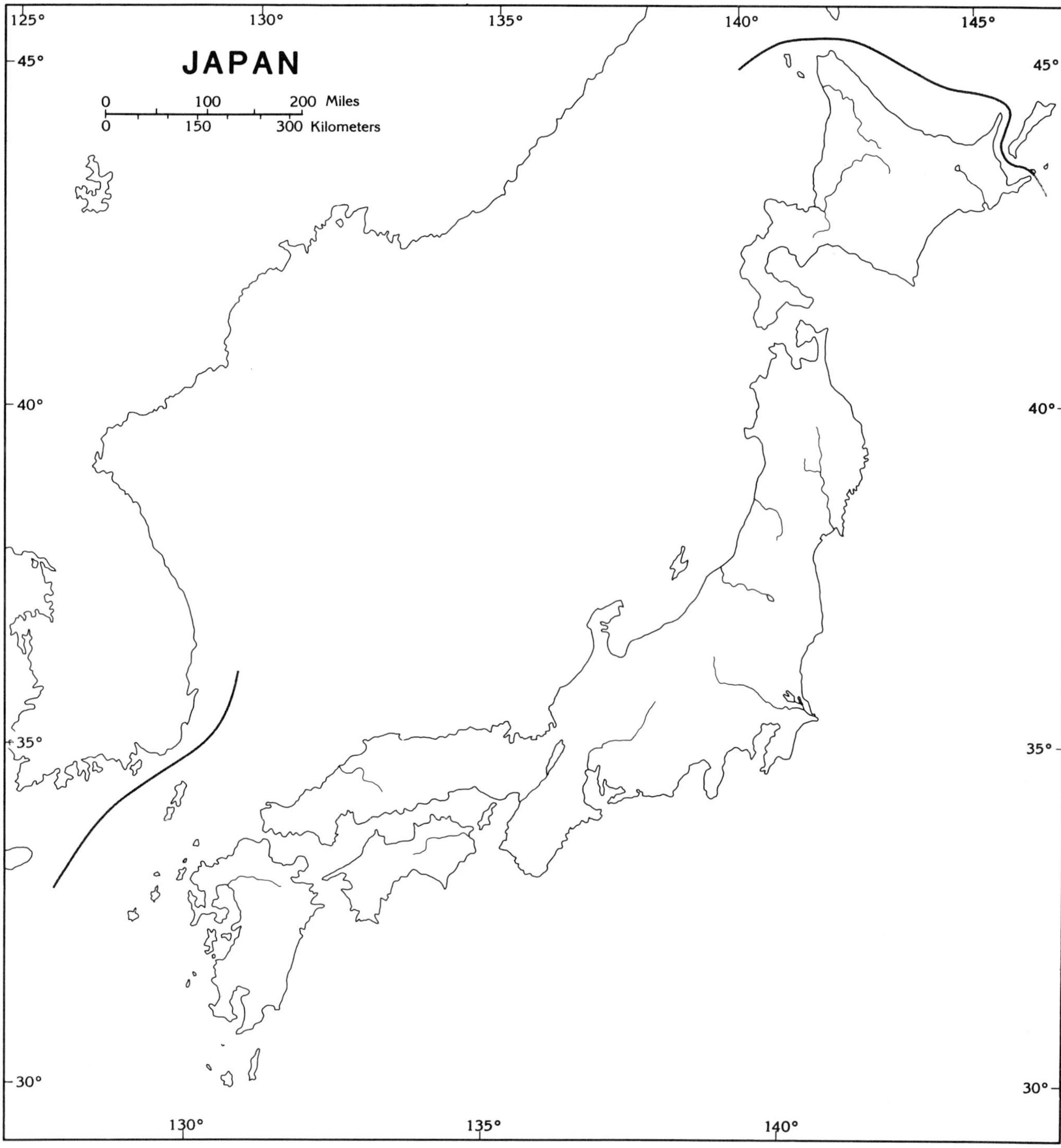
JAPAN
0 100 200 Miles
0 150 300 Kilometers
125°
130°
135°
140°
145°
45°
40°
35°
30°

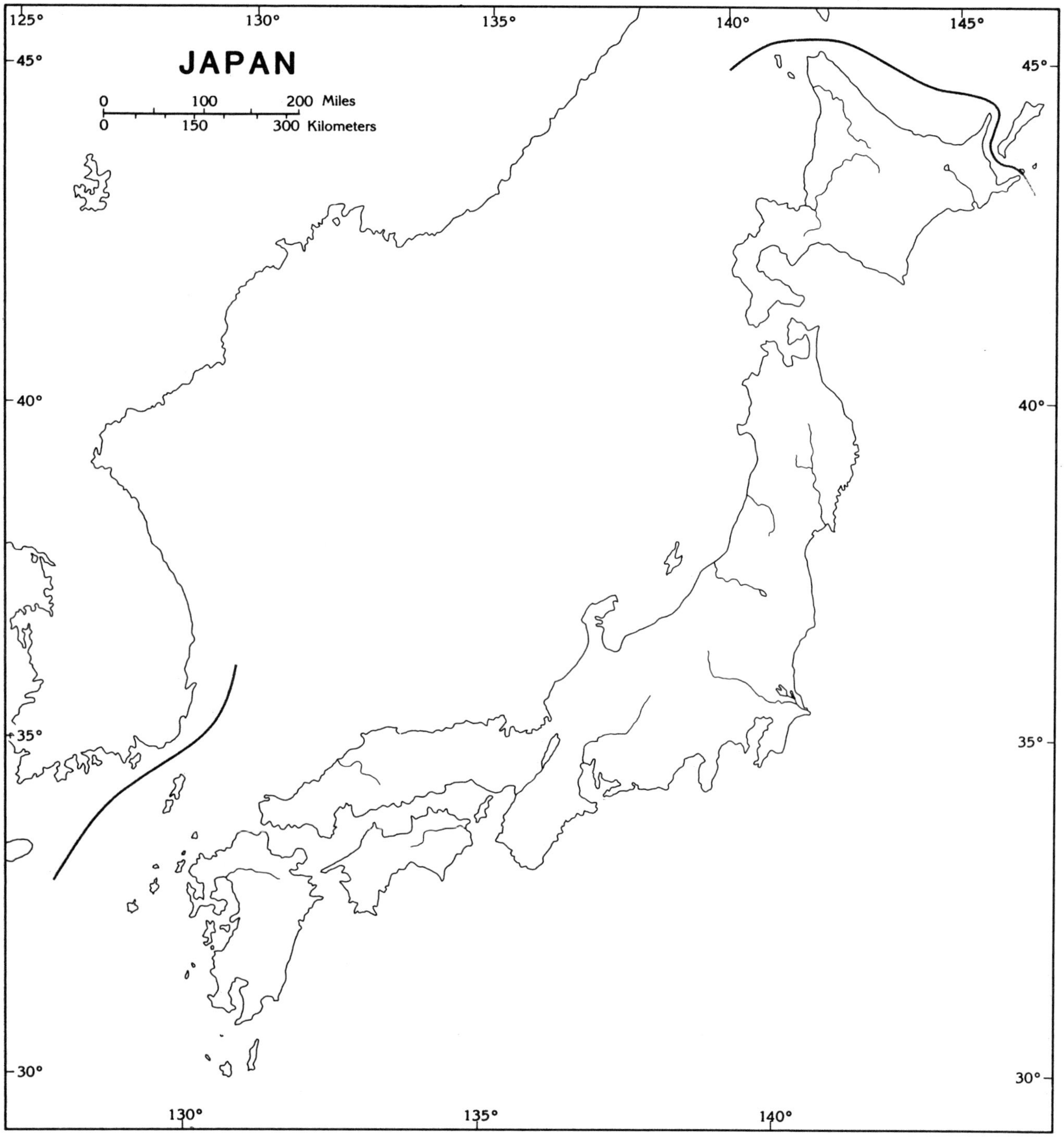

JAPAN
0 100 200 Miles
0 150 300 Kilometers
125°
130°
135°
140°
145°
45°
40°
35°
30°

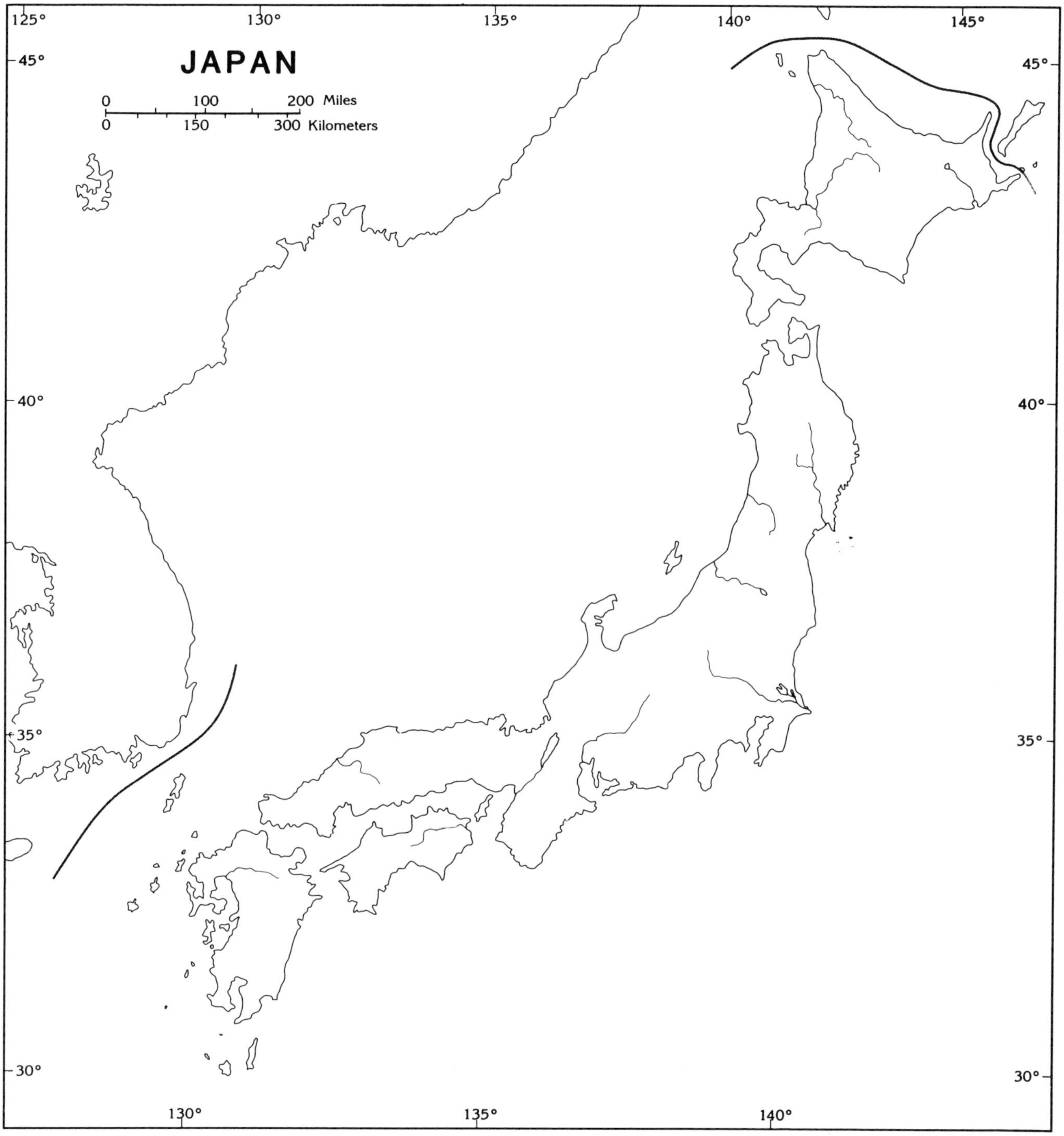
JAPAN
0 100 200 Miles
0 150 300 Kilometers
125°
130°
135°
140°
145°
45°
40°
35°
30°

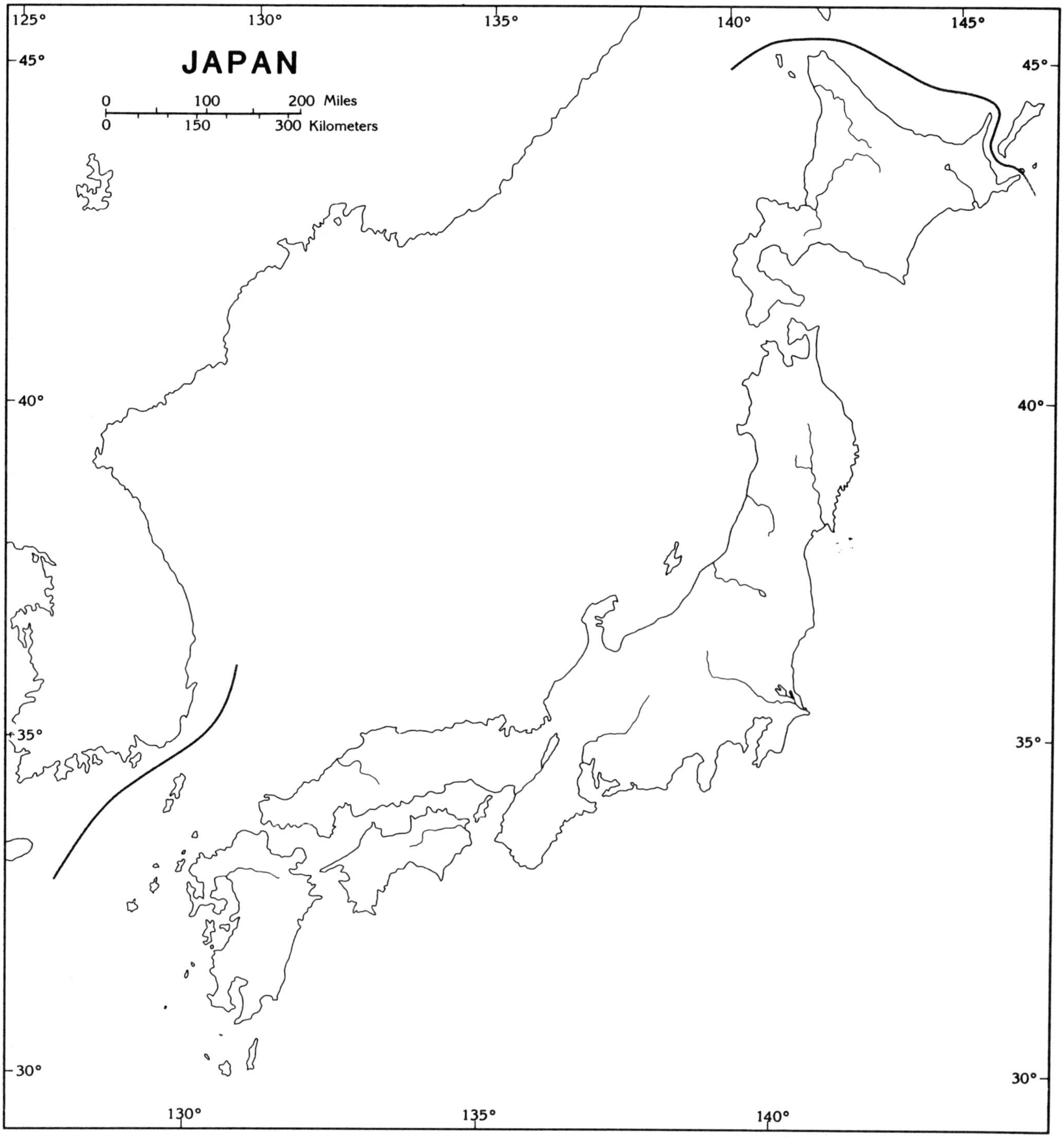

JAPAN
0 100 200 Miles
0 150 300 Kilometers
125°
130°
135°
140°
145°
45°
40°
35°
30°

CHAPTER 4: MIDDLE AMERICA--COLLISION OF CULTURES

OBJECTIVES OF THIS CHAPTER

Chapter 4 opens Part 2 of the book--the underdeveloped realms--with a survey of Middle America (Mexico, Central America, and the Caribbean islands). Historical geography is the focus of Systematic Essay 4, which also connects this subdiscipline to the rich cultural heritage of the realm. Following an introduction, the sequential influences of the Mesoamerican Indian civilizations and the Hispanic colonists are reviewed, and their various cultural collisions are evaluated as shapers of contemporary society. The regional structure of the realm is first cast within the useful "Mainland-Rimland" framework (placed against the complex European colonial legacy). Each major region is then treated: Mexico--highlighting economic geography and the intensifying urban crisis, the Caribbean--underscoring cultural geography and the impact of tourism, and Central America--emphasizing spatial dimensions of the ongoing political and military conflicts that plague these republics.

Having learned the regional geography of Middle America, you should be able to:

1. Describe the major contributions of the Mayans, Aztecs, and Spanish in shaping the contemporary cultural and social geography of the realm (and the African and other European colonial infusions in much of the Rimland portion of Middle America).

2. Understand the general workings and concerns of historical geography, and its importance in interpreting contemporary Middle America.

3. Describe the basic idea of environmental determinism.

4. Differentiate between Mainland and Rimland Middle America in terms of political, cultural, and economic regional geography.

5. Understand the geographic patterns of the Caribbean region, their evolution, and their prospects for future change.

6. Understand the geography of Mexico, especially its resource-development opportunities and worrisome problems related to out-of-control population and urban growth.

7. Understand the altitudinal zonation of environments that mark the economic and population geographies of Middle and South America.

8. Understand the geographical parameters of Central America--one of the most pivotal areas on earth in the 1980s--including the current military-political status of each of its republics.

9. Locate the major physical, cultural, and economic-spatial features of the realm on an outline map.

GLOSSARY

Land bridge (269-270)

A relatively low-lying strip of land that connects two larger landmasses, such as the 3750-mile-long (6000-kilometer-long) mainland of Middle America that connects North and South America between the U.S. border and the southeastern end of the Panamanian isthmus.

Isthmus (269)

Performs the same function as a land bridge, but is usually shorter in length and narrower in width. On the Mainland Middle American land bridge, the South American end is the isthmus of Panama; some geographers would extend the isthmus portion of Central America to cover not only the entire state of Panama (and the short protrusion of Colombia to its southeast) but much of Costa Rica as well.

Simulation model (272)

A modeling method seeking to replicate real-world spatial processes by treating them as "rules" that control the actions of "players" in a "game," with each "round" that is played producing a new spatial pattern. By adjusting the rules to produce more and more realistic outcomes, the model-builder comes closer and closer to matching reality (whose outcomes are already known); through such matching, a researcher is able to offer more meaningful forecasts about the future.

Time-geography (273)

A method for investigating human spatial behavior by measuring and analyzing the allocations of time to perform various tasks in geographic space.

Settlement geography (273)

The study of facilities people construct on the landscape while occupying a region; because such facilities are durable, they often survive beyond the time period of their initial function and thereby provide the historical geographer with valuable evidence about the past. Settlement analysis links cultural, historical, rural, and urban geography, and provides an important exercise in the field training of a geographer.

Mesoamerica (274)

Anthropological label for the Middle American culture hearth, mapped in detail in Fig. 4-4 (p. 276) and in world context in Fig. 6-4 (p. 368).

Environmental determinism (274)

The view that the natural environment has a controlling influence over various aspects of human life, including cultural development. Also referred to as environmentalism.

Mexica Empire (276-277)

The name the ancient Aztecs gave to the domain over which they held hegemony on the north-central mainland of Middle America.

Acculturation (279)

Cultural modification resulting from interculture borrowing. In cultural geography, the term is used to designate the change that occurs in the culture of indigenous peoples when contact is made with a technologically more advanced society.

Placering (280)

The initial stage of Spanish mining in the New World, during the early 16th century; entails the "washing" of gold from streams carrying gold dust and nuggets.

Greater Antilles (281 box)

The larger islands of the northern Caribbean, including Cuba, Hispaniola, Puerto Rico, and Jamaica.

Lesser Antilles (281 box)

The smaller-island arc of the eastern Caribbean, stretching southward from the Virgin Islands to Trinidad near the South American coast.

Mainland/Rimland Classification (282-283)

John Augelli's framework that recognizes a Euro-Indian Mainland and a Euro-African Rimland in Middle America, as mapped in Fig. 4-9 on p. 283.

Hacienda (283-284)

Literally, a large estate in a Spanish-speaking country. Sometimes equated with plantation, but there are important differences between these two types of agricultural enterprise and rural land tenure.

Plantation (284)

A large estate owned by an individual, family, or corporation and organized to produce a cash crop. Almost all plantations were established within the tropics; since 1950, many have been divided into smaller holdings or reorganized as cooperatives.

Ejido (284)

Communally-owned cooperative farmlands in central and southern Mexico, originating during the 1910 Revolution that saw the carving up of the haciendas.

Mulatto (289)

A person of mixed African (black) and European (white) ancestry.

Mestizo (289)

A person of mixed white and American Indian ancestry.

Transculturation (291)

Cultural borrowing that occurs when different cultures of approximately equal complexity and technological level come into close contact. In acculturation, by contrast, an indigenous society's culture is modified by contact with a technologically more developed society.

Tierra caliente (292 box)

The lowest of three vertical zones into which the topography of Middle and South America is divided according to elevation. The caliente is the hot humid coastal plain and adjacent slopes up to 2500 feet (750 meters) above sea level. The natural vegetation is the dense and luxuriant tropical rainforest; the crops are bananas, sugar, cacao, and rice in the lower areas and coffee, tobacco, and corn along the somewhat higher slopes.

Tierra templada (292 box)

The intermediate altitudinal zone in Middle and South America, between 2500 and 6000 feet (750 and 1850 meters). This is the "temperate" zone, with moderate temperatures compared to the tierra caliente. Crops include tobacco, coffee, corn, and some wheat.

Tierra fria (292 box)

The highest-lying zone below the snow line in Middle and South America, from about 6000 feet (1850 meters) up to nearly 12,000 feet (3600 meters). Coniferous trees stand here; upward they change into scrub and grassland. There are also important pastures within the fria, and wheat can be cultivated. Several major population clusters in the region lie at these altitudes. Above the fria lie the paramos, the highest and coldest zone, sometimes called tierra helada.

Insurgent state (297)

Territorial embodiment of a successful guerrilla movement. Antigovernment insurgents establishing a territorial base in which they exercise full control--thus, a "state within a state."

SELF-TESTING QUESTIONS

Cover this side of the page with a sheet of paper. Uncover each line after you have attempted to answer the question in the left column. If necessary, refer to textbook page(s) listed at the right.

Question	Answer	Page
MIDDLE AMERICAN CHARACTERISTICS	MIDDLE AMERICAN CHARACTERISTICS	
What is the spatial extent of this realm?	Mexico and Central America from the U.S. border to the northern edge of South America plus all of the Caribbean islands to the east.	269
How is Middle America more culturally diverse than South America?	African and Asian ancestries prevail beside those of European background; the Indian cultural contribution is greater; the Caribbean is a region of especially complex cultural pluralism.	269
What is a land bridge?	A narrow overland link connecting two larger landmasses.	269-70
HISTORICAL GEOGRAPHY	HISTORICAL GEOGRAPHY	
What is historical geography?	The interpretation of spatial changes on the earth's surface.	272
What is a spatial process?	A causal force that acts and unfolds over time to shape a spatial pattern.	272
What is settlement geography?	The study of the facilities people build as they inhabit a region.	273
Why are Spanish New World towns laid out in the same form?	This pattern was decreed by the Laws of the Indies, and became a tradition that was adhered to everywhere the Spanish went.	273
Why was the gridiron street pattern so effective in colonial Middle America?	It facilitated tight control over local Indians who were being converted to Christianity; if trouble broke out, soldiers could easily seal off the afflicted blocks.	273-74

MESOAMERICAN LEGACY	MESOAMERICAN LEGACY	
What is Mesoamerica?	The Middle American culture hearth that stretched from northern Mexico southeast to Panama.	274
What is the difference between Central and Middle America?	Central America is the mainland between Mexico and South America; Middle America also includes Mexico and the Caribbean region.	275 box
What is environmental determinism?	The notion that the natural environment dominates the course of cultural development.	274
Where was the Maya civilization centered?	In Guatemala and the Yucatan Peninsula of Mexico.	275
Describe some of the Mayas' accomplishments.	Urbanization, pyramids, spectacular palaces, stone carvings and other artwork, mathematics, astronomy, calendrics, advanced agriculture, and a wide trading network.	275
Who were the Aztecs?	The successors to the Toltecs, who succeeded the Mayas.	276-77
What was the Mexica Empire?	The Aztec state centered in the Valley of Mexico, headquartered at the city of Tenochtitlan.	276-77
Which crops did the Mesoamerican Indians contribute to the world?	Corn (maize), various kinds of beans, the sweet potato, the tomato, cacao, squash, and tobacco.	277-78
What kinds of agriculture were emphasized after the Spanish conquest?	The keeping of livestock, especially cattle and sheep--which competed with the growing of the subsistence crops of the conquered Indians.	278-79
What was the most far-reaching change in the cultural landscape brought by the Spanish?	The resettlement of the Indians from rural land into villages and towns laid out and controlled by the conquerors; these towns were administrative centers for tax collection and labor recruitment, especially for mining.	279
What further changes were engendered by the mining industry?	The creation of a network of tightly-controlled Spanish towns that cemented power over even the more isolated parts of Mexico.	280-81

MAINLAND AND RIMLAND	MAINLAND AND RIMLAND	
What is the difference between the cultures of Mainland and Rimland Middle America?	The Mainland is of Euro-Indian character; the Rimland is dominated by a Euro-African cultural heritage.	282-83
What are the geographical differences between Mainland and Rimland?	The Rimland was an area of sugar and banana plantations, high accessibility, seaward exposure, and maximum cultural contact and mixture; the Mainland was removed from these contacts, the region of the <u>hacienda</u>, more self-sufficient, and less dependent on outside markets.	283-84
What are the five characteristics of Middle American plantations?	They are located in the tropical coastlands and islands; they produce single export crops; foreign ownership and profit outflow dominate; labor is seasonal and has often been imported; "factory in the field" methods are far more efficient than those of the <u>hacienda</u>.	284
What are the political legacies of Middle American colonialism?	The Mainland states are now independent republics, but all except Belize have Hispanic origins; the cultural variety of the Caribbean is much greater--Cuba, Puerto Rico, and the Dominican Republic were Spanish; Haiti and several islands of the Lesser Antilles were French; Jamaica, Trinidad, and many lesser islands were British; and other small islands had Dutch, Danish, and U.S. affiliations.	284-85

THE CARIBBEAN	THE CARIBBEAN	
Who were the indigenous peoples of the pre-European period?	The Arawaks and Caribs, who were soon decimated.	281 box
Why do the export crops and the mineral resources of the islands provide so modest an income for the region, and why does widespread poverty persist?	These commodities face severe competition from many other Third World countries, and are not established on a scale that could improve local living standards; most islanders therefore live in poverty, eking out a subsistence existence from a small plot of poor land.	287
Why is the Caribbean region mostly a legacy of Africa?	Much of its population has descended from slaves that were brought from Africa to work the plantations; black and <u>mulatto</u> populations heavily dominate most of the islands.	288-89
What is a <u>mestizo</u>?	A person of mixed white-Indian ancestry.	289
Which islands have sizable Asian populations?	Trinidad, Jamaica, Guadeloupe, and Martinique.	289

MEXICO	MEXICO	
How strong is the Indian imprint on Mexican culture?	Extremely strong: 60 percent of the population are mestizos, 30 percent are pure Indian, and only 9 percent are European.	291
Did the 1910 Revolution achieve its goals?	For the most part, yes: the _haciendas_ were redistributed, many into communally-owned _ejidos_; the revolution also resurrected the Indian cultural contribution to Mexican life.	291-92
Name the four altitudinal zones of human activity.	_Tierra caliente_ (sea level to 2500 feet), _tierra templada_ (2500-6000 feet), _tierra fria_ (6000-12,000 feet), and _tierra helada_ (above 12,000 feet).	292 box
Where is most of Mexico's oil production located?	Along the central and southern Gulf Coast, especially around the city of Villahermosa and the nearby Bay of Campeche.	293

CENTRAL AMERICA	CENTRAL AMERICA	
Name the seven republics of Central America.	Guatemala, Honduras, Belize, El Salvador, Nicaragua, Costa Rica, and Panama.	295
Where is most of the region's population concentrated?	In the _templada_ zone of the highlands, toward the Pacific side of the land bridge.	295
In which republic is the U.S. aiding the government against a Marxist-led insurgency?	El Salvador.	301
Which republic is run by the Sandinistas, and which republic is the "Switzerland" of Central America?	Nicaragua; Costa Rica.	302-3
What is the current status of the Panama Canal?	The U.S. withdrawal is proceeding, with the Canal itself to be turned over to Panama by 2000.	304

MAP EXERCISES

MAP COMPARISON

1. The map of Middle America's regions (Fig. 4-2, pp. 270-271) clearly offsets each of the three components of this realm. Compare and contrast the environments of these regions, basing your observations on the world maps of landscapes (Fig. I-5), precipitation (Fig. I-9), climates (I-10), vegetation (Fig. I-12), and soils (Fig. I-13) in the Introduction chapter (pp. 10-25).

2. Discuss the population distribution of mainland Middle America since 200 A.D. by comparing the map of Mesoamerican historical geography (Fig. 4-4, p. 276) with the appropriate portion of the contemporary world population map (Fig. I-14, p. 26).

3. Compare Figs. 4-8 and 4-9 (pp. 282, 283) and make observations about the colonial influence on the Mainland/Rimland split. Why is the Central American east coast included in the Rimland rather than the Mainland region? What generalizations can be made about the composition of present Caribbean populations based on the European countries that once ruled them?

MAP CONSTRUCTION

(Use outline maps at the end of this chapter)

1. In order to familiarize yourself with Middle American physical geography, place the following on the first outline map:

   a. Water bodies: Caribbean Sea, Gulf of Mexico, Bay of Campeche, Gulf of California, Gulf of Tehuantepec, Gulf of Honduras, Gulf of Panama, Gulf of Darien, Gulf of Fonseca, Panama Canal, Rio Grande River, Lake Nicaragua, Straits of Florida, Windward Passage

   b. Land bodies: Bahama Islands, Greater Antilles, Lesser Antilles, Leeward Islands, Windward Islands, Virgin Islands, Baja California, Yucatan Peninsula, Isthmus of Panama, Hispaniola, Florida Keys, Sierra Madre Occidental, Sierra Madre Oriental, Valley of Mexico, Sonora Desert, Isthmus of Tehuantepec

2. On the second map, enter the name of every country that is shown, including all of the mainland republics and as many island-states as possible.

3. On the third outline map, urban-economic information should be entered as follows:

   a. Capital cities (locate and label with the symbol *): Mexico City, Belmopan, Guatemala City, San Salvador, Tegucigalpa, Managua, San Jose, Panama, Havana, Kingston, Nassau, Port-au-Prince, Santo Domingo, San Juan, Fort-de-France, Port of Spain, Willemstad

   b. Other cities (locate and label with the symbol ·): Ciudad Juarez, Tijuana, Monterrey, Torreon, Chihuahua, Durango, Zacatecas, Guadalajara, Tampico, Veracruz, Acapulco, Oaxaca, Villahermosa, Merida, Cozumel, Belize City, Quetzaltenango, San Pedro Sula, San Miguel, Leon, Bluefields, Limon, Colon, Santiago de Cuba, Mariel, Guantanamo, Montego Bay, Puerto Plata, Mayaguez

c. Economic Regions (identify with circled letter):

A - Curacao
B - Major area of ejidos today
C - Bay of Campeche oilfield
D - Panama Canal
E - Mexican gold placer area

PRACTICE EXAMINATION

SHORT ANSWER QUESTIONS

Multiple Choice

1. The racial term applied to people of mixed white and Indian ancestry is:

   a) mulatto
   b) caliente
   c) ejido
   d) mestizo
   e) americano

2. In which mainland republic did the Sandinistas overthrow the Somoza regime in the late 1970s:

   a) Nicaragua
   b) El Salvador
   c) Honduras
   d) Dominican Republic
   e) Mexico

3. The "rifles-and-beans" policy is associated with which of the following countries:

   a) Trinidad
   b) Cuba
   c) Mexico
   d) Puerto Rico
   e) Guatemala

True-False

1. The core area of the Aztec state was located in what is still the core area of Mexico today.

2. The large island of Trinidad is located in the Greater Antilles.

3. Costa Rica is a U.S. enemy that is a haven for Marxist insurgencies throughout Middle and South America.

Fill Ins

1. The gridiron street pattern was first introduced to Middle America by Europeans from the country of ______________.

2. The lowest-lying altitudinal zone of agricultural activity, extending from sea level to an elevation of 2500 feet (750 meters), is the tierra ______________.

3. The Yucatan Peninsula is a part of the country of ______________.

Matching Question on Middle American Countries

_____	1.	U.S. backs government in civil war	A.	Guatemala
_____	2.	U.S. backs "Contra" rebels	B.	Puerto Rico
_____	3.	"Switzerland" of Central America	C.	Dominican Republic
_____	4.	Large East Indian population	D.	Curacao
_____	5.	Eastern half of Hispaniola	E.	Jamaica
_____	6.	Former Danish colony	F.	U.S. Virgin Islands
_____	7.	Autonomous U.S. commonwealth	G.	Honduras
_____	8.	Largest Caribbean city	H.	Haiti
_____	9.	Bauxite mining	I.	Belize
_____	10.	Formerly belonged to Colombia	J.	Panama
_____	11.	Still a Dutch colony	K.	Nicaragua
_____	12.	Western half of Hispaniola	L.	Costa Rica
_____	13.	Capital is Tegucigalpa	M.	Cuba
_____	14.	Maya culture hearth	N.	Trinidad
_____	15.	Formerly British Honduras	O.	El Salvador

ESSAY QUESTIONS

1. One of this realm's most important regional frameworks is the Mainland/Rimland scheme. Compare and contrast the physical, economic, and cultural geographies of each component of Middle America, and show how its historical evolution proceeded in a direction that will probably prevent regional integration and unity in the future.

2. Discuss the European impact on the shaping of the cultural and political geographies of the Caribbean region today. Why is this region trapped in a cycle of poverty that is rooted in the economic system inherited from the colonial era?

3. Describe the fourfold vertical zonation of environments in "Latin" America, including the agricultural geography of each zone. How has the low-latitude population distribution of this part of the world accommodated itself to the environmental zonation?

4. Few countries in the underdeveloped world possess the opportunities and challenges facing Mexico. Discuss those aspects of Mexican geography that offer a real potential for meaningful progress in living standards in the foreseeable future, and weigh them against the problems the country confronts in the short- and longer-term future.

5. Much of Central America's current turmoil is rooted in this region's past. Trace the historical geography of the region, paying attention to the Indian, Spanish, and other cultural influences. How has the economic system of the past few centuries contributed to the present upheaval?

TERM PAPER POINTERS

The "Term Paper Pointers" section of Chapter 1 in this Study Guide offered suggestions about approaching research and writing on geographic realms and their components, and you may wish to consult this material if you are undertaking a report on a Middle American region.

The only geography of this entire realm is West & Augelli, a splendid survey that is practically indispensable for doing a report on Middle America. Four more recent, but general, overviews of both Middle and South America may prove helpful too--Blakemore & Smith, Blakemore et al., Blouet & Blouet, and James & Minkel. For regions within the realm, see Eyre, Feinberg, Lowenthal, MacPherson, Miller & Miller, Mintz & Price, Mitchell, and Rumney. Individual countries are treated in Augelli [1986], Bolland Browning, Calvert, Christian, Dominguez, Floyd, Halebsky & Kirk, Hall, Hamilton & Harding, Jamaica, Levy & Szekely, Peckenham & Street, Pico, Riding, Sherlock, Tata, Tata & Lee, and both Walker sources. The political turmoil of this realm is covered in Anderson, Crowley & Griffin, Driever, and McColl. A good background on the Third World as a whole is Dickenson et al., and a classic work on tropical environments is Gourou. Several historical and/or anthropological titles have been included to cover the Meso-american legacy: Benjamin & McNellie, Davidson & Parsons, Driver, Helms, McCullough,

Nietschmann, Turner, Wagley & Harris, Wauchope, Weaver, and Wolf. The colonial legacy is treated in Dozier, Grant, Lovell, McAlister, and Sanderson. On special topics, regionalization is treated by Augelli [1962], migration by Brown & Shue, Jones, and Richardson, environmental determinism in Huntington and Martin, Mexican oil in Falk, tourism in Sealey, deforestation in Caufield, National Research Council, and Myers, and resource management in Wilken. The increasingly important topic of urbanization is covered in Griffin & Ford, Hope, Sargent, Scott, Stanislawski, and Wilkie; the cover story in Time of August 6, 1984 is an updating of this topic, with pp. 26-35 focused on Mexico City. A good set of geographical case studies can be found in Boehm & Visser. Three useful atlas materials are Arbingast, Diagram Group, and Wilkie.

The Systematic Essay on historical geography is another fruitful place from which to develop a term paper, and if you do you may want to locate two important periodicals that many libraries have--the Journal of Historical Geography and the Historical Geography Newsletter put out by the specialty-group of that name which is affiliated with the Association of American Geographers. Classical statements on this subdiscipline are offered by Clark and Sauer, with updates by Jakle and Norton; some of the newer approaches to the subject are found in Carlstein et al. and Pred. The fascinating subject of landscape interpretation is treated in Meinig, with a fine overview by Lewis; the periodical Landscape should also be consulted. Finally, you are reminded that the North America chapter-end bibliography (pp. 246-247) contained notable titles: Brown, Lewis [1983c], Mitchell & Groves, Meinig [1986], and Ward [1979].

SPECIAL EXERCISE

The "Latin" American city is an increasingly important focus of attention in the contemporary geography of Middle and South America. The generalization developed by Griffin & Ford (pp. 324-325) is a useful model of intraurban spatial structure that can be used to interpret current developments, and this exercise will explore that linkage.

Begin by reading the entire article by Griffin & Ford which appeared in the Geographical Review in 1980; a review of that article and a number of other examples can be found in the later chapter by these authors that appeared in the 1983 book, Cities of the World (both Griffin-Ford references are listed on text p. 344). You will then need to locate two recent articles describing life in Mexico City: (1) "A Proud Capital's Distress," Time, August 6, 1984, 26-35; and (2) "Mexico City: An Alarming Giant," National Geographic, August 1984, 138-185. Once you have found and read these articles, provide answers to the following questions:

1. Discuss the application of the Griffin-Ford model to Mexico City, citing the author's own ideas about this connection.

2. How might Griffin and Ford have adjusted their model to Mexico City as this metropolis evolved over the past 400 years? Would a time-stage model along the lines of Adams (text pp. 215-216) be helpful here, and what differences might be noted?

3. In the <u>Time</u> article, which land-use zone is shown in the picture on p. 26? On p. 28? In the 3 pictures on pp. 34-35? Make a photocopy of the map on p. 30, and then superimpose the Griffin-Ford zones as carefully as you can.

4. In the <u>National Geographic</u> article, how does the Griffin-Ford model relate to the maps on pp. 146-147 (make a sketch map of the application)? In the pull-out photograph on pp. 176-179, which zone(s) are we looking at?

5. Briefly discuss the future of Mexico City and its spatial structuring by linking the Griffin-Ford model to the material presented in the box on pp. 293-294. Might an alternative spatial pattern help Mexico City to cope with its intensifying problems.?

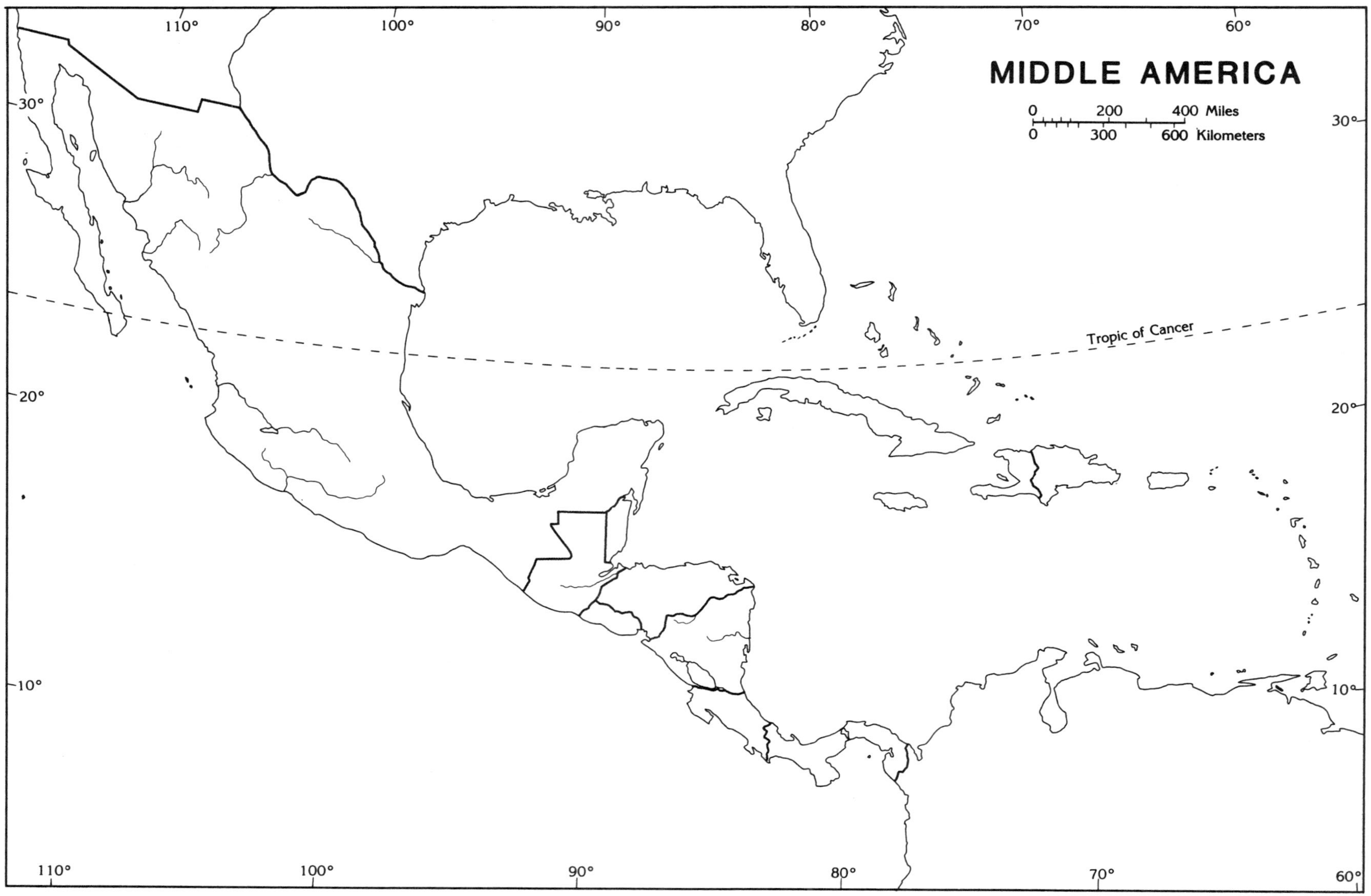
MIDDLE AMERICA
0 200 400 Miles
0 300 600 Kilometers
Tropic of Cancer
110°
100°
90°
80°
70°
60°
30°
20°
10°

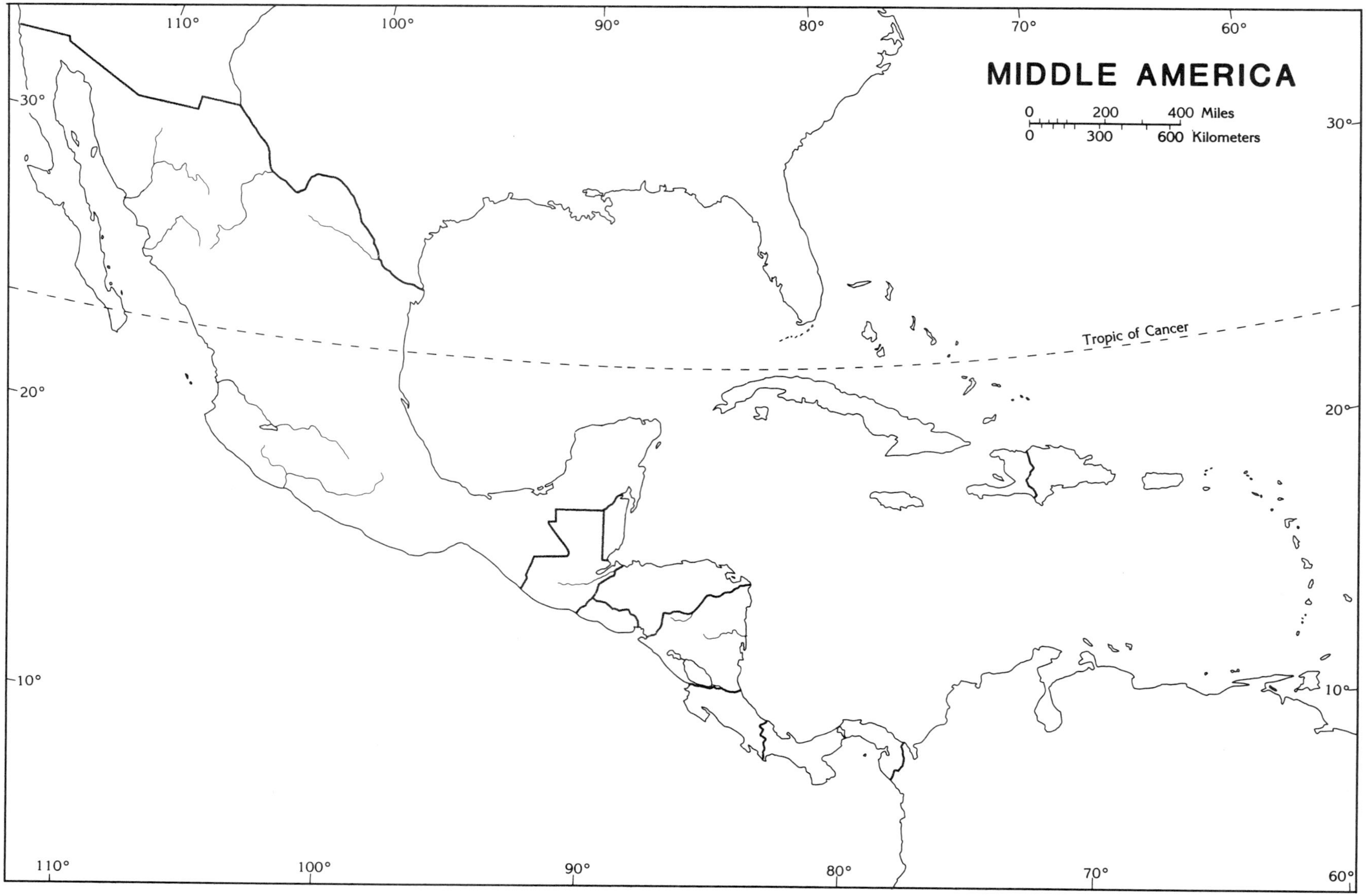
MIDDLE AMERICA
0 200 400 Miles
0 300 600 Kilometers
Tropic of Cancer
110°
100°
90°
80°
70°
60°
30°
20°
10°

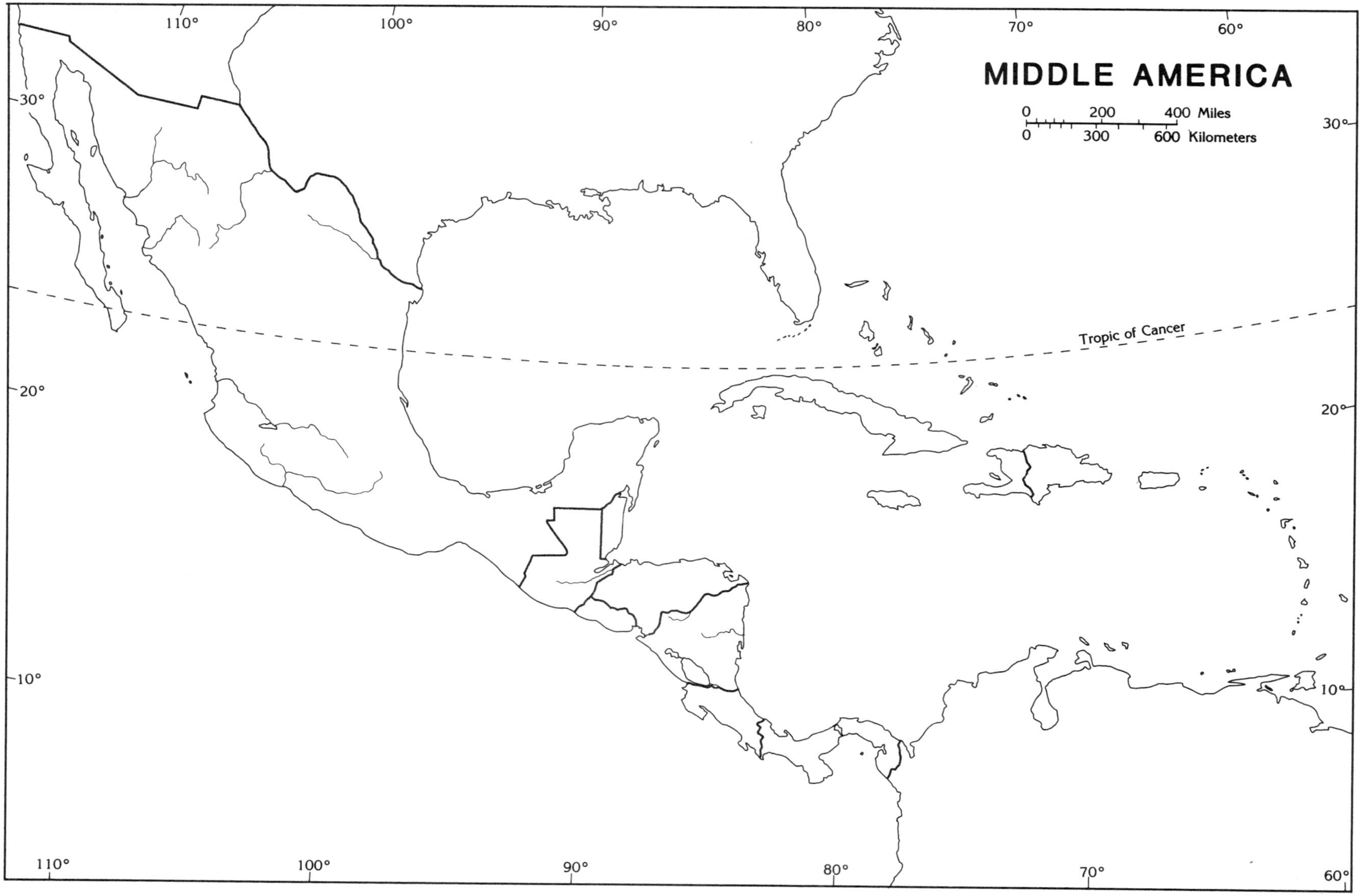
MIDDLE AMERICA
0 200 400 Miles
0 300 600 Kilometers
Tropic of Cancer
110°
100°
90°
80°
70°
60°
30°
20°
10°

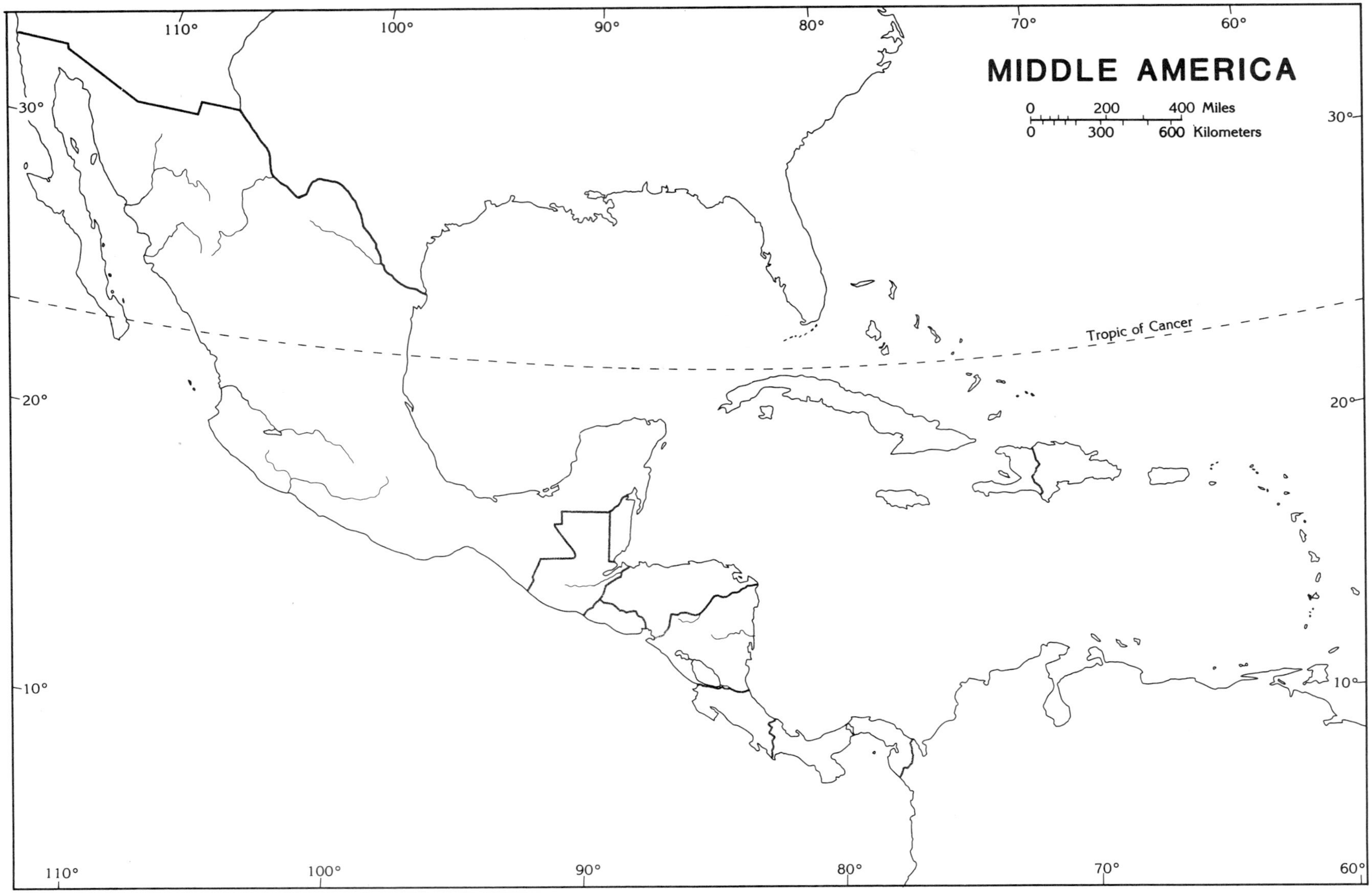
MIDDLE AMERICA
0 200 400 Miles
0 300 600 Kilometers
Tropic of Cancer
110°
100°
90°
80°
70°
60°
30°
20°
10°

CHAPTER 5: SOUTH AMERICA--TRADITION AND TRANSITION

OBJECTIVES OF THIS CHAPTER

Chapter 5 treats South America, a realm of vast potential but also of widespread frustrations in the effort to move itself forward. The Systematic Essay--economic geography--focuses on agriculture, a dilemma for most of the underdeveloped world. Following a general background, South America's historical geography is traced in some detail, with emphasis on its emerging culture areas. Urbanization is introduced, emphasizing a model of city structure that may fit most Third World realms. A review of South America's regions completes the chapter, presented within a three-part framework involving the Caribbean North, the Andean West, and the Mid-latitude South; Brazil is treated separately in a Vignette that immediately follows this chapter (both here and in the text).

Having learned the regional geography of South America, you should be able to:

1. Understand the general distribution of world agricultural systems, with particular emphasis on the spatial pattern of food production in the South American realm.
2. Grasp the essentials of the historical geography of this continent.
3. Understand the various influences that have shaped South America's culture areas.
4. Deepen your understanding of Middle and South American urbanization, building on the Griffin-Ford model introduced in Chapter 4 on pp. 324-325.
5. Describe the broad regionalization pattern of the realm, with reference to cultural patterns, historical, and political developments.
6. Understand the geographic essentials of each South American republic, including its resources, natural environments, and economic development potential.
7. Locate the leading physical, cultural, and economic-spatial features of the realm on an outline map.

GLOSSARY

Components of the spatial economy (312)

The four sets of productive economic activities that were elaborated on text p. 226: the primary, secondary, tertiary, and quaternary sectors of the economy in their various spatial expressions.

Agriculture (312)

The purposeful tending of crops and livestock, expressed in the world's major agricultural systems that are mapped in Fig. 5-3 on pp. 312-313.

Subsistence agriculture (313-314)

The opposite of commercial agriculture in which crop and livestock surpluses are sold for profit; by subsistence is meant an existence based upon only the bare necessities to maintain life--spending most of one's time in pursuit of enough food, clothing, and shelter to survive.

Altiplano (310)

High-elevation plateau or basin between even higher mountain ranges; Andean altiplanos often lie at altitudes in excess of 10,000 feet (3000 meters).

Papal Line (317)

The line mediated by the Pope codified in Treaty of Tordesillas of 1494: all Latin American territory west of 50°W longitude belonged to Spain, all land to the east was Portugal's.

Barrios (325)

The desperate slums (sometimes called favelas) of cities in Middle and South America, found in the disamenity sector as well as the zone of peripheral squatter settlement (the photograph on p. 326 gives a good idea of these conditions as well as the incredible social contrast with nearby affluent areas).

Model of the Latin American city (324-325)

The Griffin-Ford model of Latin American intraurban spatial structure, discussed and diagrammed in the box on pp. 324-325, which may apply to the Third World city in general.

Plaza (324-325)

The old hub and focus of the Latin American city, the open central square flanked by the main church and government buildings (see photo p. 325).

Llanos (327)

Savannalike grasslands, especially of the low-lying Orinoco Basin in Venezuela as well as neighboring Colombia and Guyana.

Oriente (333)

Literally "the east," refers to the jungly lowlands of Peru and Ecuador to the east of the Andes that are sparsely populated but contain petroleum deposits.

OPEC (334)

The Organization of Petroleum Exporting Countries, the international oil cartel whose membership is mapped in Fig. 6-13.

Pampa (337)

Literally the word means "plain," the physiographic subregion that is Argentina's leading crop-and-livestock-producing region.

SELF-TESTING QUESTIONS

Cover this side of the page with a sheet of paper. Uncover each line after you have attempted to answer the question in the left column. If necessary, refer to textbook page(s) listed at right.

Question	Answer	Page
SOUTH AMERICAN CHARACTERISTICS	SOUTH AMERICAN CHARACTERISTICS	
How does South America's longitudinal position differ from North America's?	It lies considerably farther to the east, closer to Africa but facing a much wider Pacific Ocean on the west.	309
What is the realm's dominant physiographic feature?	The Andes Mountains, which form an imposing north-south barrier along the realm's entire west coast.	309
ECONOMIC GEOGRAPHY	ECONOMIC GEOGRAPHY	
What is economic geography?	The study of the diverse ways in which people earn a livelihood, and how the goods and services they produce are spatially organized.	312
What forms the center in a global-scale Von Thunen model?	The "world city" would be the European and U.S. edges of the North Atlantic Basin, suggested in Fig. 1-5 on p. 63.	312
HISTORICAL GEOGRAPHY	HISTORICAL GEOGRAPHY	
Who were the Incas?	Descendants of ancient peoples who created a major civilization in the northern Andes around A.D. 1200, centered in Peru's Cuzco Basin.	310
List some of the Incas' major achievements.	Most of all, the political integration of Andean South America; a splendid circulation system for goods and ideas; administration of complex social and economic system; Quechua--a regional language that still survives.	314-15
How and when did the Spanish conquer the Incas?	In 1534, Pizarro led a small band of soldiers into Cuzco and deposed the Inca Empire which was ripe for an overthrow given its internal divisions and problems of royal succession.	315

How did Spain and Portugal resolve their 15th century disputes over South American territory?	The 1494 Treaty of Tordesillas, mediated by the pope, gave Spain all land west of the 50th meridian and Portugal all territory to the east.	317
Why did the Portuguese import so many Africans to coastal Brazil?	They opted for a Caribbean-style plantation economy, which required the services of millions of African slaves.	317-18
How did South America's 19th century independence movement spread?	Argentina and Chile, farthest from Peru, began it and Bolivar pushed down from New Granada to the north--by 1824, the Spanish were driven out; in Brazil, the Prince Regent's son defied his father, proclaimed independence, and expelled the loyalist Portuguese forces in 1822.	319

CULTURE AREAS

CULTURE AREAS

How well unified is the South American realm?	Surprisingly little interaction occurs among countries, which are often themselves fragmented into pluralistic societies of great complexity.	320
Name the 5 internal culture regions of this realm as defined by Augelli.	*Tropical plantation* region of the northeast and northern coasts; *European-commercial* region of most of the mid-latitude south; the *Indo--subsistence* region of the Andean valleys and plateaus; the *Mestizo-transitional* region covering large parts of the interior between Indian and commericial agriculture; and the *Undifferentiated* primitive areas of the most isolated zone (largely Amazonia).	320-22

URBANIZATION

URBANIZATION

How fast has South American urbanization increased recently?	Very rapidly: in 1925, the realm was 33 percent urban, by 1950 40 percent, but over 60 percent in 1975 and about 75 percent today; the urban areas now increase annually by 5 percent, the rural areas by only 1.5 percent.	322
Why does the migration toward cities remain so high in the 1980s?	*Push* factors are rural poverty and lack of land reform; *pull* factors are perceived employment opportunities, education, medical care, and a more exciting pace of life.	323-24
Name the land-use zones of the "Latin" American city.	The CBD, the commercial spine, the elite residential sector, the zone of maturity, the zone of *in situ* accretion, the zone of peripheral squatter settlements, and the disamenity sector containing the *favelas*.	324-25

CARIBBEAN SOUTH AMERICA	CARIBBEAN SOUTH AMERICA	
What forces bind this region?	Common coastal location; a tropical plantation culture and economy; large black and Asian minorities.	326
What are Venezuela's major activities?	Lake Maracaibo oil, iron ores in the east, and the developing llanos cattle and farming.	327-28
What are the political statuses of the Guianas?	Guyana is independent, with unstable leadership; Suriname, also unstable, is now independent of the Dutch; French Guiana remains an overseas department of France.	328 box
Where is most of Colombia's coffee grown?	In the tierra templada zone of the Andean slopes, centered in Antioquia-Caldas Provinces.	329
THE ANDEAN WEST	THE ANDEAN WEST	
Which states comprise this region? Which culture dominates? Which groups rule?	Peru, Ecuador, Bolivia, and Paraguay; the Indo-subsistence cultural sphere is dominant, with a large mestizo population besides the majority Indians: nonetheless, the European elite holds most of the political power.	330
What is the sequence of subregions proceeding across Peru and Ecuador from west to east?	The fairly dry coastal plain; the high Andes, interspersed with altiplanos; the interior Andes slopes of rainforest-covered Montana; the Amazon Basin jungles (Oriente) which contain oilfields.	330-32
What is Bolivia's leading export commodity by value?	Tin, mined from some of the world's richest deposits; oil and natural gas are becoming ever more important.	335
MID-LATITUDE SOUTH AMERICA	MID-LATITUDE SOUTH AMERICA	
What is the significance of the Argentine Pampa?	Productive meat-and-grain region which has made the country a major food exporter; gave Buenos Aires a rich hinterland, spurring its growth and success.	337-38
What was the War of the Pacific all about?	Bolivia's long search for an outlet to the sea led to a conflict with Chile (1879-1884) over a corridor across the northern Atacama Desert --which proved to contain valuable mineral deposits.	336 box

Compare and contrast the population distributions of Argentina, Chile and Uruguay.	Argentina has a strongly peripheral pattern; Chile has a highly concentrated distribution focused on Middle Chile; Uruguay's population is uniformly spread across its territory.	337-40
Compare and contrast the shapes of Chile and Uruguay.	Chile's is extremely elongated, causing potential political problems; Uruguay's is compact, easy to govern and manage.	339-40
What are northern Chile's major minerals?	Nitrates in the Atacama; copper too, especially in the vicinity of Chuquicamata.	341
What is the status of the Falkland ("Malvina") Islands in the mid-1980s?	The British rule with an iron hand following their 1982 ouster of the Argentinian military; negotiations will have to await a calmer time in the future, but the prospects for an amicable solution look unpromising.	340-41 box

MAP EXERCISES

MAP COMPARISON

1. Since this is the first chapter to treat a realm located in the southern hemisphere, this is an appropriate place to compare environmental distributions north and south of the equator. Start with the Hypothetical Continent (Fig. I-11, p. 20): describe the similarities and differences, and to what extent the southern hemisphere "mirrors" the northern in its climatic regionalization. What additional details are provided by the empirical climate map (Fig. I-10, pp. 18-19)? How do the hemispheric distributions of vegetation (Fig. I-12, pp. 22-23) and soils (Fig. I-13, pp. 24-25) relate to each other and to the climatic pattern?

2. The final sentence in column 2 on text p. 309 suggests in words the impact of the Andes on the realm's physical geography. Compare the maps referred to, setting forth your observations in some detail. Reread the box on altitudinal zonation (p. 292), and discuss the application of this scheme to the Andes in the country of Peru.

3. In the underdeveloped realms agriculture is especially closely associated with the biophysical environment. Accordingly, compare the broad agricultural patterns of the world map (Fig. 5-3, pp. 312-313) with the global distributions of precipitation, climate, vegetation, and soil (Figs. I-9 through I-13, pp. 14-25). What correspondences can be discerned? How do areas of commercial agriculture differ from those of subsistence agriculture in the relative "favorability" of local environments?

4. South America's complex cultural geography often affects the ability of a state to integrate and effectively govern its citizens. After carefully studying Fig. 5-8 (p. 321), rank the realm's countries according to their apparent cultural uniformity--assuming that the more homogeneous cultures are the stronger nation-states. Provide a brief justification for each country, but qualify it if you have learned appropriate additional information that might affect overall unity (or lack thereof).

5. On Fig. 5-15 (text p. 331), rule a straight line in light pencil from Callao-Lima on the coast to Iquitos in the northeastern Oriente; draw an identical line on the physiographic map (Fig. 5-2, p. 311). On a blank piece of paper draw a cross-section of this traverse (using both maps to assist you), showing the general shape of the surface involved (your instructor can assist you in doing this). Then, using Fig. 5-8 (p. 321), add in the cultural association of each segment of the cross-section, and describe its leading characteristics according to the text discussion (pp. 320-322).

MAP CONSTRUCTION

(Use outline maps at the end of this chapter)

1. In order to familiarize yourself with South American physical geography, place the following on the first outline map:

    a. Rivers: Amazon, Orinoco, Paraguay, Parana, Uruguay, Sao Francisco, Magdalena, Cauca, Apure, Maranon, Madeira, Guayas, Rio de la Plata, Colorado, Xingu, Tocantins, Negro

    b. Water bodies: Amazon Delta, Lake Maracaibo, Lake Titicaca, Gulf of Guayaquil, Strait of Magellan, Plata Estuary, Humboldt (Peru) Current

    c. Land bodies: Cape Horn, Tierra del Fuego, Falkland Islands, Chiloe Island, Guiana Highlands, Mato Grosso Planalto, Andean Altiplano, Patagonian Plateau, Pampa, Chaco, Llanos, Brazilian Highlands, Atacama Desert, Andes Mountains

2. On the second map, political-cultural information should be entered as follows:

    a. Label each country, and its capital (*)

    b. Reproduce the cultural map (Fig. 5-8, p. 321) using an appropriate color-pencil scheme

3. On the third outline map, urban-economic information should be entered as follows:

    a. Cities (locate and label with the symbol •): Caracas, Ciudad Bolivar, Georgetown, Paramaribo, Cayenne, Valencia, Barquisimeto, Santa Marta, Medellin, Bogota, Buenaventura, Quito, Esmeraldas, Guayaquil, Lima, Callao, Iquitos, Huancayo, Arequipa, La Paz, Oruro, Potosi, Asuncion, Arica, Antofagasta, Chuquicamata, Valparaiso, Santiago, Concepcion, Valdivia, Montevideo, Buenos Aires, Rosario, Cordoba, Mendoza, Tucuman, Punta Arenas, Sao Paulo, Rio de Janeiro, Santos, Porto Alegre, Belo Horizonte, Curitiba, Brasilia, Volta Redonda, Salvador, Recife, Belem, Fortaleza, Manaos

b. Economic regions (identify with circled letter):

A - Itaipu Dam
B - Pampas
C - Antioquia-Caldas
D - Oriente
E - Magdalena Valley
F - Guayas Lowland
G - Titicaca Basin
H - Middle Chile
I - Minas Gerais
J - Grande Carajas Scheme

**(n.b.: Brazil [text pp. 346-359] is included in the above exercises because there are no maps in the Vignette that follows this Study Guide chapter).

PRACTICE EXAMINATION

SHORT-ANSWER QUESTIONS

Multiple Choice

1. South America's largest city in population size is:

   a) Mexico City
   b) Sao Paulo
   c) Buenos Aires
   d) Rio de Janeiro
   e) Caracas

2. The landlocked country whose search for an outlet to the sea launched the "War of the Pacific" was:

   a) Bolivia
   b) Panama
   c) Brazil
   d) Paraguay
   e) Amazonia

3. Which of the following regions is not located in Argentina:

   a) Patagonia
   b) Pampa
   c) Maracaibo Lowland
   d) Chaco
   e) Entre Rios

True-False

1. Suriname is a former colony of France, and today enjoys complete independence.

2. Uruguay is a classic example of an <u>elongated state</u>.

3. Itaipu Dam is situated on the border between Brazil and Venezuela.

Fill Ins

1. The two leading resources of Chile's Atacama Desert are nitrates and __________.

2. At the center of the "Latin" American City model, one finds a land-use zone called the __________________.

3. The coffee-growing areas of Colombia are concentrated in the elevation zone called <u>tierra</u> ______________.

Matching Question on South American Countries

_____	1.	Still under colonial rule	A.	Colombia
_____	2.	Former Dutch Guiana	B.	Venezuela
_____	3.	Contains Cape Horn	C.	Guyana
_____	4.	Cuzco Basin	D.	Suriname
_____	5.	Contains realm's largest city	E.	French Guiana
_____	6.	East side of Plata estuary	F.	Ecuador
_____	7.	Victor of "War of the Pacific"	G.	Peru
_____	8.	Contains two capitals	H.	Bolivia
_____	9.	Lower Orinoco Basin	I.	Paraguay
_____	10.	Population more than 50 percent Asian	J.	Chile
_____	11.	Site of Chibcha civilization	K.	Argentina
_____	12.	West bank of Itaipu Dam	L.	Uruguay
_____	13.	Guayas Lowland	M.	Brazil

ESSAY QUESTIONS

1. The Argentinian writer-philosopher Borges likened the 1982 Falklands War with Britain to "a fight between two bald men over a comb." Discuss the claims made by each country for their remote island territory (actually an archipelago), and the possibilities for an amicable solution. How strategic a location is the Falkland chain? Could the relative importance of its regional position change in the future?

2. Discuss the altitudinal zonation scheme of "Latin" American environments [introduced in box on text p. 292] as they apply to the Andean highlands of northwestern South America, highlighting the distinctive cultural and economic geographies associated with each zone.

3. Discuss the historical geography of Peru over the past 1000 years. Why did great advances occur here in the pre-European period? Why did the Spanish headquarter their colonial empire here after 1535? And why has Peru ceased to be one of South America's leading countries in this century?

4. Discuss the agricultural geography of South America. Draw a sketch map showing the commercial, subsistence, and non-agricultural areas of the realm. Why did this particular pattern emerge, and what relationship does it bear to the cultural spatial differentiation of the continent?

5. The economic geography of northern and western South America has in part been shaped by the exploitation of major mineral resources. Discuss the spatial patterns of production that characterize the following metallic and fossil-fuel resources: oil and natural gas, nitrates, copper, silver, and iron ore. What key minerals are missing or are unfavorably distributed in the drive to modernize these countries?

TERM PAPER POINTERS

The "Term Paper Pointers" section of Chapter 1 in this Study Guide offered suggestions about approaching research and writing on geographic realms and their components, and you may wish to consult this material if you are undertaking a report on a South American region.

South America's diversity again necessitates that one obtain a deeper background from a good regional geography. The best work is James & Minkel, but this book is a shadow of its earlier editions; the 4th edition by James alone remains a classic, but it is outdated in many of the subjects it covers. Several other dated works may be worth consulting as well in your library: Butland, Cole, Gilbert, and Platt. Of the more recent works (Blakemore & Smith, Bromley & Bromley, and the two books by Morris) none compares to James, but they are useful because of their currency and because they contain good bibliographies of the geographical literature of the past few years (an important feature for studying a realm that is undergoing such swift change). Regional-level works are offered by Black, Caviedes, and Gauhar; individual countries are covered by Fifer, Lombardi, Mertz & Myers, Gorman, Lewis, Scobie, and Hendelman. Another overview is Blouet & Blouet, a smaller volume arranged topically rather than regionally, which contains valuable overviews of "Latin" American environments, historical geography, transportation, agriculture, population characteristics, urbanization, mining and manufacturing, and international contexts. A similar book, though based on shorter case studies, is Boehm & Visser. Aguilar and J. Wilkie present country profiles and statistical portraits that are updated annually. Miller & Miller have compiled a still-useful realwide bibliography, and much information is available in Blakemore et al.

Several cross-matching topical works are worth noting too, many from disciplines outside geography. Cultural and historical geography are treated in Augelli, Bailey & Nasatir, Denevan [1976], Harris, Hennessy, Morner, Parsons, Steward & Faron, and Wagley. Political geography is covered by Braveboy, Child, Goebel, Super & Wright, and the "Falklands" article. Issues of economic development are highlighted in Enders & Mattione, Jumper, Maos, Odell & Preston, Preston, and Sanchez-Albornoz. Agriculture is treated in Barraclough, Denevan [1985], Dahlberg, Grigg [all sources], Ilbery, Manshard, Preston, Rotberg & Rabb, and Rumney. Urbanization trends are elucidated by Armstrong & McGee, Bourne, Drakakis-Smith, Gilbert & Gugler, Gilbert & Ward, Gilbert et al., Griffin & Ford [both sources], Gwynne, Hauser, Lowder, Linn, Roberts, and R. Wilkie; Weil examines recent work in medical geography. A delightful travel book on "Latin" America by train is Theroux, a pithy but perceptive observer of the ever-changing scene; he has also written The Great Railway Bazaar about a similar journey across Eurasia as well as The Kingdom by the Sea, a survey of the coastline of the United Kingdom. The geography of cocaine is treated in "Cocaine Wars," Hudson, and Stearman.

The Systematic Essay on economic geography can also lead to a number of research papers, but the subject is so broad it is necessary to first choose among such still-large topics as agriculture, manufacturing, trade, and services. Recent general works to consult for further literary pointers are Jumper et al., Wheeler & Muller, Berry et al., and Wilbanks; other cited works on text pp. 344-345 with South American connections are Boehm & Visser, Blouet & Blouet, Cole, Grigg [1974], Morris [1981], Odell & Preston, and Taylor. The closely related subject of economic development--mentioned above--is also explored further in Systematic Essay 9 (text pp. 518-520) and good references can be pursued from the bibliography on pp. 564-565: Chisholm, Cole, de Souza & Porter, Dickenson, Forbes, Gore, Johnson, Kresge, Mabogunje, Muegge & Stohr, Reitsma & Kleinpenning, Rostow, Simpson, and Vining.

SPECIAL EXERCISE

The geography of energy is a fascinating and increasingly important pursuit in economic geography. As a first step in learning more about world spatial patterns of energy resources, carefully study the map on text pp. 378-379 and its accompanying graphs on p. 379. Once you have become familiar with this distribution, answer the following questions in some depth.

1. Provide an overview of the global correspondence between energy reserves and population distribution (use maps on pp. 26-29), highlighting those regions and countries that are best and worst off.

2. Review each of the realms covered so far (Europe, Australasia, Soviet Union, North America, Japan, and Middle America), and offer a brief assessment of its relative strength; focus especially on the United States and the Soviet Union, discussing their domestic fuel-resource bases and the longer term prospects for each superpower.

3. For South America, offer a detailed evaluation of its prospects and problems as to the future course of economic development; what if oil were to be found in quantity (as many suspect) in the waters of the Falkland Islands--how could political problems affect the economic geography of this remote corner of the world?

4. Carefully contrast the differences in production and consumption as shown in the graphs on p. 379; extrapolating from that information, predict the flows of international trade in energy (heavily dominated by petroleum) which result and check those flows against a current map in your library. (See a recent atlas or current publications of the U.S. Department of Energy).

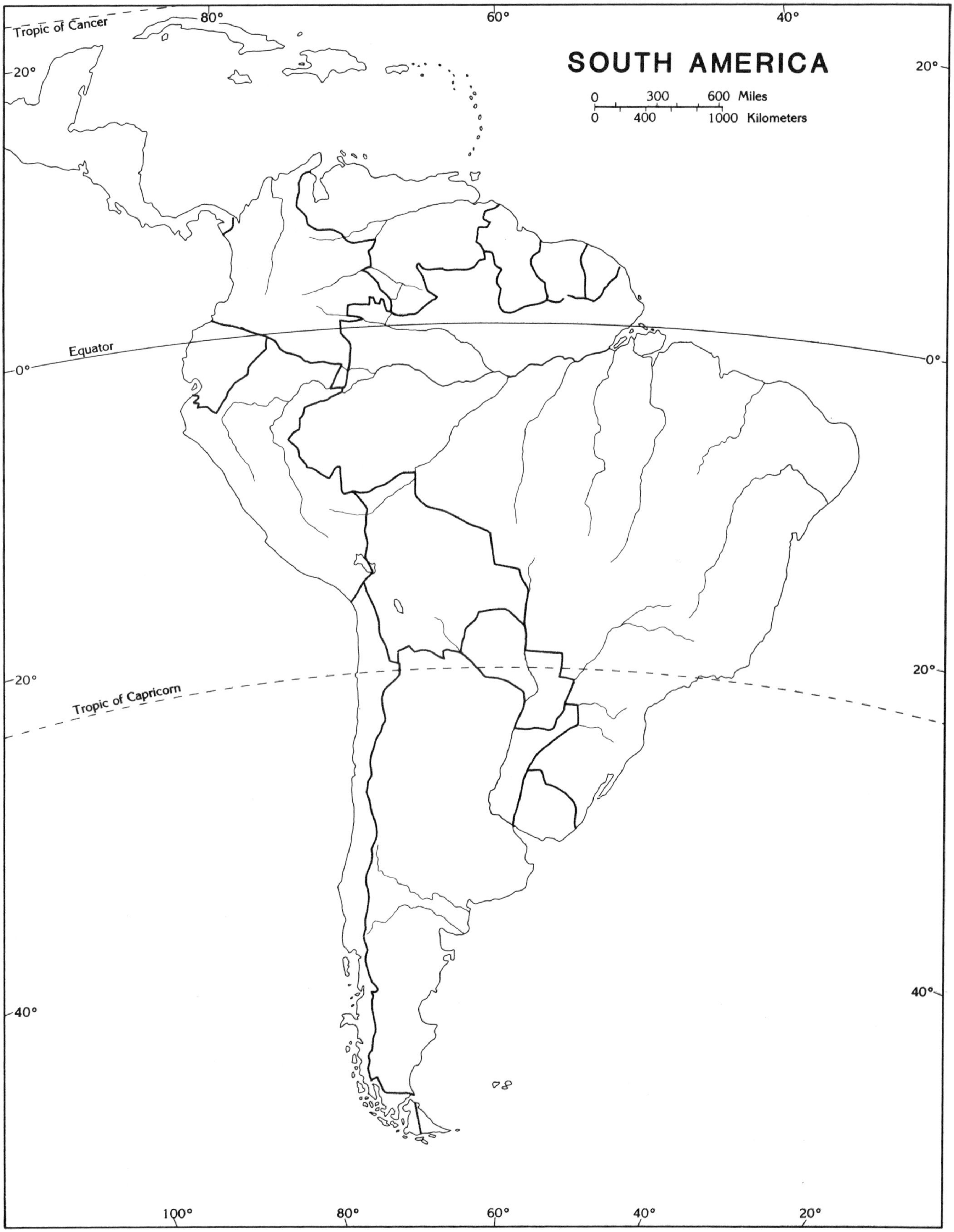
SOUTH AMERICA
0 300 600 Miles
0 400 1000 Kilometers
Tropic of Cancer
Equator
Tropic of Capricorn
80°
60°
40°
20°
0°
20°
40°
100°
80°
60°
40°
20°

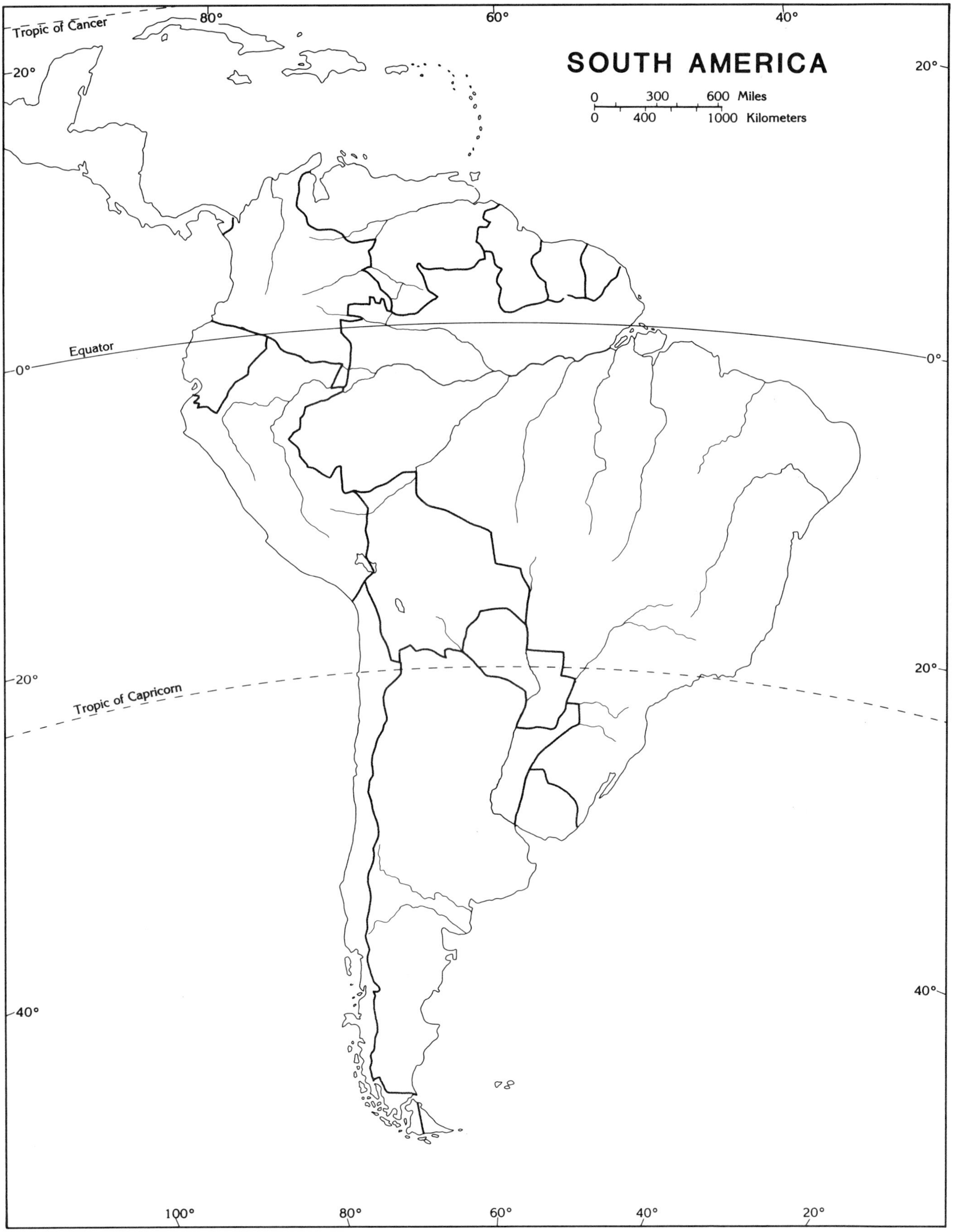
SOUTH AMERICA
0 300 600 Miles
0 400 1000 Kilometers
Tropic of Cancer
Equator
Tropic of Capricorn
80°
60°
40°
20°
20°
0°
0°
20°
20°
40°
40°
100°
80°
60°
40°
20°

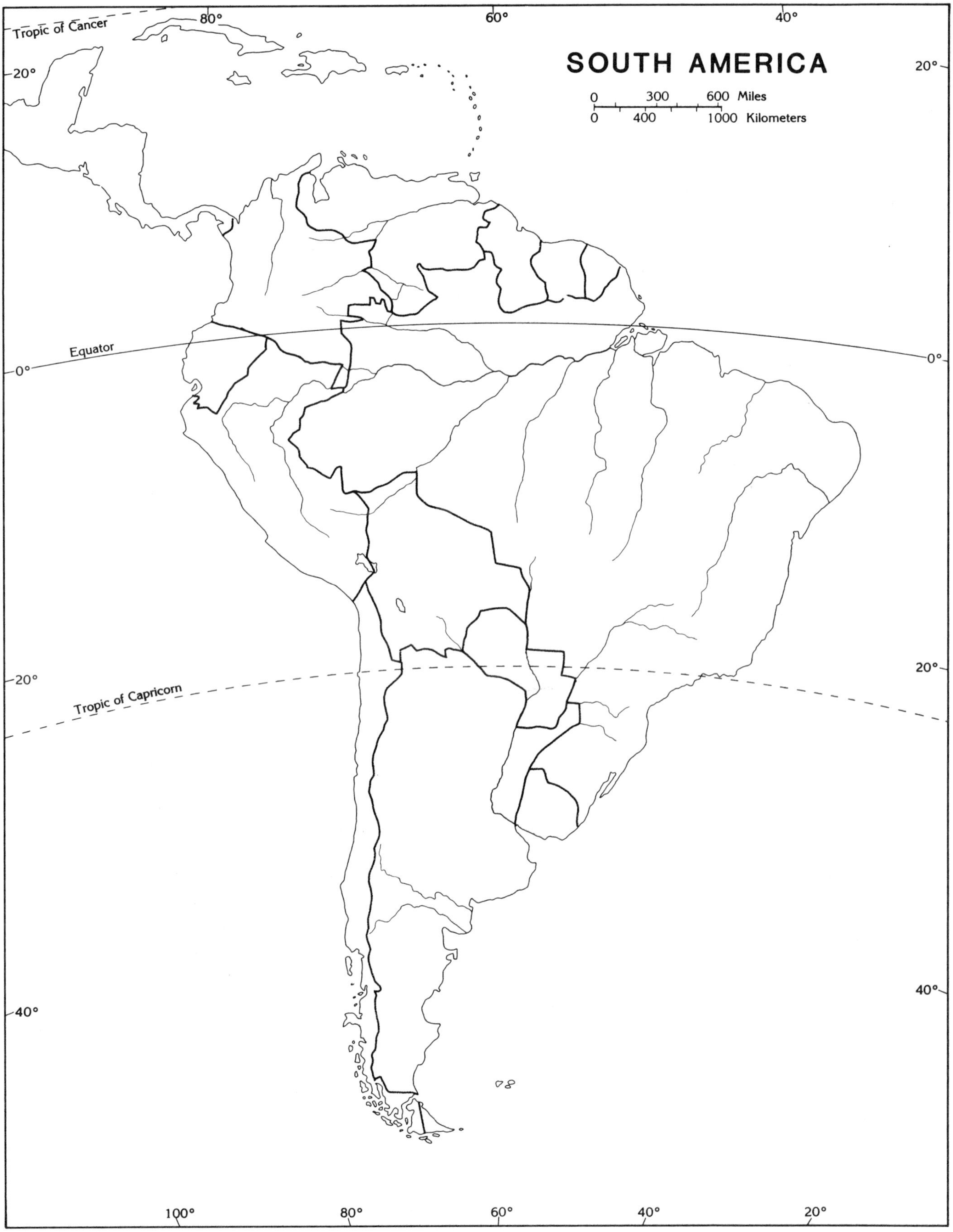

SOUTH AMERICA
0 300 600 Miles
0 400 1000 Kilometers
Tropic of Cancer
Equator
Tropic of Capricorn
80°
60°
40°
20°
20°
0°
0°
20°
20°
40°
40°
100°
80°
60°
40°
20°

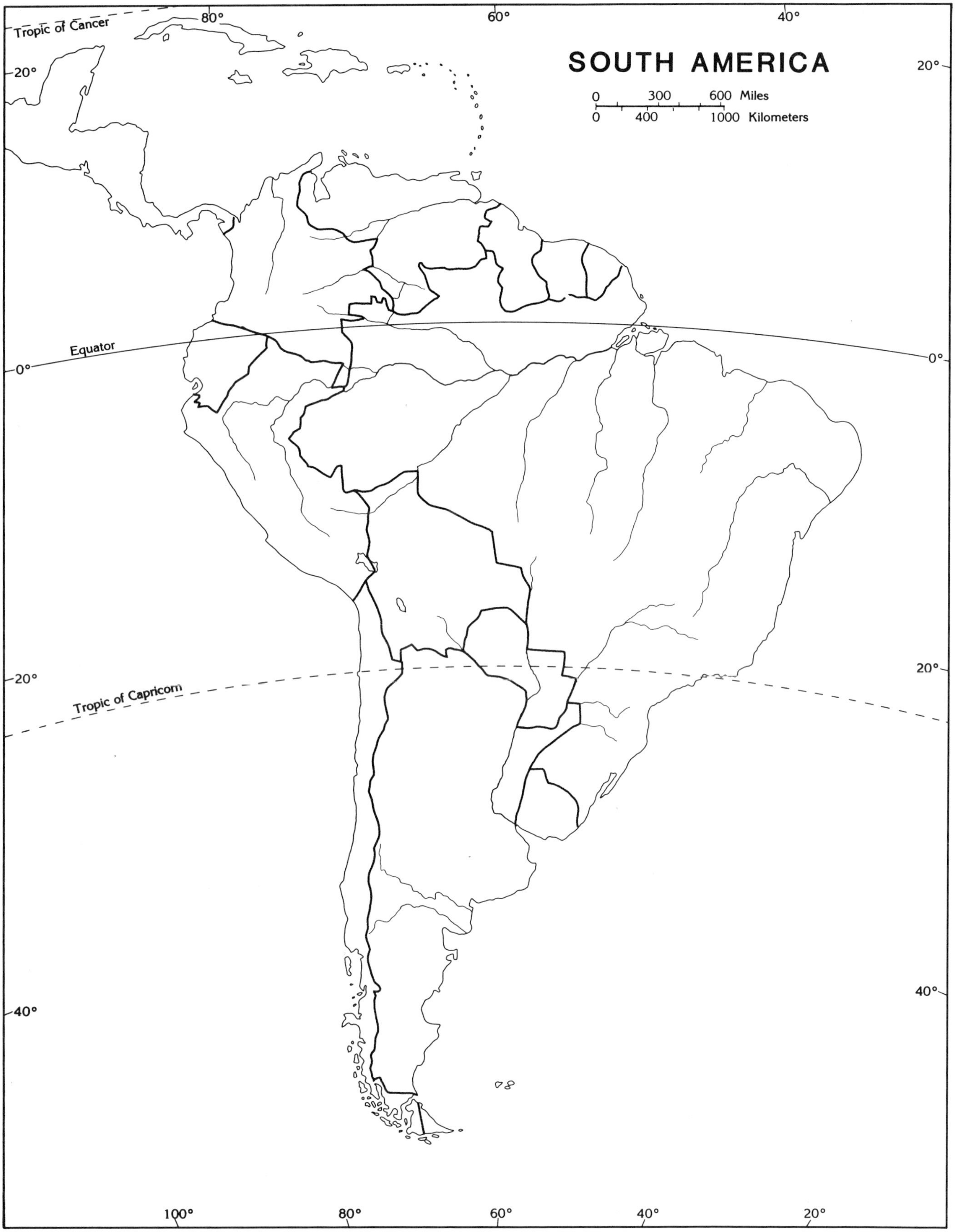

SOUTH AMERICA
0 300 600 Miles
0 400 1000 Kilometers
Tropic of Cancer
Equator
Tropic of Capricorn
80°
60°
40°
20°
20°
0°
0°
20°
20°
40°
40°
100°
80°
60°
40°
20°

VIGNETTE: EMERGING BRAZIL--POTENTIALS AND PROBLEMS

OBJECTIVES OF THIS VIGNETTE

This Vignette covers Brazil, a country and a region of South America that by virtue of its population and territorial size stands apart--and merits separate treatment. Moreover, unlike its continental neighbors, Brazil is at the threshold of an economic takeoff that may rank it among the world's major powers sometime in the next century; but it must first overcome some awesome problems which have been exacerbated by a foreign-debt crisis that arose suddenly in the 1980s. The Vignette begins with a brief introduction that leads to survey of the country's six major regions--with the opportunities and challenges facing each region highlighted. Brazil's severe demographic problems are then examined, and the Vignette concludes with an overview of development problems that considers internal and external forces.

Having learned the regional geography of Brazil, you should be able to:

1. Appreciate its place in South America and on the current world scene.
2. Describe its gargantuan geographic dimensions.
3. Understand its regional structuring.
4. Describe the contents of, and differences among, the Northeast, Southeast, Sao Paulo, South, Interior, and North regions.
5. Understand Brazil's population patterns and why they pose a great problem.
6. Understand Brazil's modernization and development thrust, the role of growth poles, and the problems that must be overcome to achieve a lasting economic takeoff.
7. Locate the leading physical, cultural, and economic-spatial features of this country on an outline map (a skill already tested in the preceding chapter of this Study Guide).

GLOSSARY

Tropic of Capricorn (346)

Latitude $23^1/2^o$ S (approx.), the parallel marking the most southerly location at which the sun's noon altitude is directly overhead (90 degrees above the horizon); the sun reaches this position each year at the winter solstice--approx. December 22--and in Brazil this latitude passes directly through metropolitan Sao Paulo.

Gasohol (347)

A gasoline substitute manufactured from sugarcane-based alcohol; in the late-1980s, over half of Brazil's automobiles use this fuel (more than in any other country), thereby greatly reducing the need for expensive petroleum imports.

Sertao (349)

The dry inland backcountry of northeastern Brazil; animal grazing dominates.

"Super Northeast" Program (349)

A multifaceted aid program, begun in the early 1980s by the federal government of Brazil, to help alleviate the miseries of a prolonged drought in the Northeast and to encourage the economic development of that troubled region.

Fazendas (350)

Large-scale coffee plantations.

Centro Oeste (351)

The region known as "the Interior," covering most of west-central Brazil; region No. 5 in Fig. B-2 (p. 348), comprising the states of Goias, Mato Grosso, and Mato Grosso do Sul as well as the capital district of Brasilia.

Forward capital (351)

A capital city deliberately located near a sensitive zone with a neighboring state or a frontier that a country wishes to develop; such a statement concerning the push to the empty interior was made in the 1950s when Brazil decided to relocateits federal headquarters from Rio de Janeiro to Brasilia.

Selva (352)

Tropical rainforest.

Grande Carajas Scheme (352)

Huge, multifacted government-business regional development program, located in the southeastern Amazon Basin.

Growth Pole (357)

An urban center with certain attributes that, if augmented by investment support, will stimulate regionwide economic development of its hinterland.

SELF-TESTING QUESTIONS

Cover this side of the page with a sheet of paper. Uncover each line after you have attempted to answer the question in the left column. If necessary, refer to textbook page(s) listed at the right.

Question	Answer	Page
BRAZIL'S CHARACTERISTICS	BRAZIL'S CHARACTERISTICS	
How large is Brazil's relative size?	Its population total of 150 million (1988) ranks it among the world's largest; territorially it is fifth biggest, after the Soviet Union, Canada, China, and the U.S.	346
What are Brazil's leading mineral resources?	Iron, bauxite, manganese, coal, and several ferroalloys; oil and gas may exist in significant quantitites, and explorations continue.	347
BRAZILIAN REGIONS	BRAZILIAN REGIONS	
What is Brazil's politico-geographical framework?	A federal republic consisting of 23 states, 3 territories, and the federal district of Brasilia.	347
What are the leading characteristics of the Northeast?	A plantation economy, poverty, overpopulation, drought; belated major economic development now under way.	347-49
What are the leading characteristics of the Southeast?	Modern Brazil's core area, largest population clusters, and a major manufacturing center.	349
What are the leading characteristics of the Sao Paulo region?	The focus of ongoing development, major coffee producer, leading area of heavy industry, and the city of Sao Paulo.	350
What are the leading characteristics of the South?	Commercial agriculture, modern technology, coal resources, and Itaipu Dam and its development potential.	351
What are the leading characteristics of the Interior?	Tropical Savannas, an active frontier, nascent development, and Brasilia.	351-52
What are the leading characteristics of the Amazonian North?	The Amazon Basin's vast potential, projects to speed growth, environmental problems, and the negative consequences of the too-rapid development of this last frontier.	352-54

DEVELOPMENT PROBLEMS	DEVELOPMENT PROBLEMS	
What population problems persist?	Internal racial divisions, and especially the persistently high birth rate.	354
How has the Brazilian economy fared since 1980?	Ups and downs; the great promise of a "takeoff" has been at least temporarily derailed by a foreign debt crisis and doubts surrounding the new civilian government leadership.	356
What is the so-called "Brazilian Model"?	A development program in which government participates in a partnership with private business--limiting the influence of foreign investors.	357
What is the growth pole concept?	A development plan in which a set of "seed" industries are nurtured, thereafter setting off "ripples" of growth in the surrounding area.	357

MAP EXERCISE

There is no map exercise in this Vignette: Brazil was included in the map activities of the previous chapter.

PRACTICE EXAMINATION

SHORT-ANSWER QUESTIONS

Multiple Choice

1. Brazil does not border:

    a) Paraguay
    b) Venezuela
    c) Chile
    d) Peru
    e) Argentina

2. Which region contains Brazil's culture hearth:

a) the Northeast
b) Sao Paulo
c) Pampas
d) the Interior
e) the North

True-False

1. Belo Horizonte and Salvador are Amazon Basin growth poles.

2. Brazil is the world's second largest Roman Catholic country.

Fill-Ins

1. Brazil's largest city is ______________.

2. The capital of Brazil before the founding of Brasilia was ______________.

Matching Question on Brazilian Cities

_____	1.	National culture hearth	A.	Rio de Janeiro
_____	2.	End of "Gold Trail"	B.	Manaos
_____	3.	Iron-ore source	C.	Brasilia
_____	4.	Forward capital	D.	Salvador
_____	5.	Rubber boomtown	E.	Volta Redonda
_____	6.	Steelmaking center	F.	Sao Paulo
_____	7.	Leading manufacturing center	G.	Lafaiete

ESSAY QUESTIONS

1. Although Brazil's progress has slowed in this decade, it still possesses vast growth potential. Discuss the advantages of Brazil's developmental opportunities, emphasizing its natural resources in a region-by-region survey.

2. Multinational corporations play a major role in the economies of most Third World countries. Discuss Brazil's experience with the multinationals over the past 20 years, and the current prospects for change in that relationship.

TERM PAPER POINTERS

The literature cited on pp. 358-359 offers a variety of materials on Brazil, but you are reminded that much is also available in the general sources on South America provided on pp. 344-345. It is also worth keeping in mind that much of Brazil's progress will depend upon events in the late 1980s--in particular, the resolution of the debt crisis and the struggle of the civilian government to maintain effective control. Therefore you should consult appropriate current sources, and your instructor and librarian will be helpful here.

There is no Special Exercise for this Vignette

CHAPTER 6: NORTH AFRICA/SOUTHWEST ASIA--FUNDAMENTALISM VERSUS MODERNIZATION

OBJECTIVES OF THIS CHAPTER

Having completed our survey of the western hemisphere's realms, we now return to the eastern hemisphere whose remaining Old World realms will occupy our attention for the rest of the book. Historically, the North African/Southwest Asian realm is properly regarded as the world culture hearth--and is the most suitable place to present the Systematic Essay on cultural geography as well as the Model Box on spatial diffusion principles. Following a brief introduction and definitional discourse on the complexity of this realm, its historical evolution is traced emphasizing the prominence of religion. Traditional culture, of course, still weighs heavily on the daily affairs of this realm, and even its vast oil production is influenced by these considerations. An in-depth regional treatment follows, highlighting such major trouble spots as Lebanon, Israel, Libya, Cyprus, Iran, and Iraq.

Having learned the regional geography of North Africa/Southwest Asia, you should be able to:

1. Appreciate the complexities involved in defining and naming this realm.

2. Understand the nature of cultural geography, its subdivisions, and its important linkages to regional geography.

3. Describe the history of this realm, stressing its role in the development of many of the world's leading religions.

4. Appreciate the significance of Islam for this realm as a whole, and the internal geographical variations of that faith.

5. Explain the major processes of spatial diffusion and be aware of the broad geographic patterns they shape.

6. Describe the world pattern of fossil fuels, emphasizing the production of petroleum in the so-called Arab World.

7. Understand the major trends within each of this realm's regions, and why so many global political problems have arisen here.

8. Describe the technique of "remote sensing" and why it has become an important geographic tool in recent years.

9. Explain the application of the Von Thunen model to the case of the Third World city.

10. Locate the major physical, cultural, and economic-spatial features of the realm on an outline map.

GLOSSARY

Popular culture (362)

The ever-changing "mass" culture of advanced societies that are open, individualistic, and class-structured.

Folk culture (362)

The strongly traditional, durable way of life found in the more isolated areas.

Cultural ecology (363)

The multiple relationships between human cultures and their natural environments.

Possibilism (363)

The doctrine of European origin that argued against environmental determinism (see text p. 274), claiming that people through their cultures are free to select from a number of environmental "possibilities."

Culture hearth (364-365)

A source area or innovation center from which cultural traditions are transmitted.

Mesopotamia (365)

The Tigris-Euphrates Plain of present-day Iraq--literally "the land between the rivers"--which is the hearth of civilization.

Fertile Crescent (365)

An arc stretching between Mesopotamia and the southeast Mediterranean coast, site of early plant domestications and farming innovations.

Spatial diffusion (372)

The geographic spreading of ideas, innovations, and information.

Relocation diffusion (372)

Diffusion by migration wherein innovations are carried by a relocating population.

Expansion diffusion (372)

The most common form of spatial diffusion wherein innovations spread within a locationally-fixed population.

Contagious diffusion (372)

Local-scale diffusion, strongly controlled by distance from the point of origin.

Hierarchical diffusion (373)

Macro-scale diffusion through an urban system, involving the "trickling down" of an innovation from atop a hierarchy to each of the lower levels in turn.

Diffusion barrier (374-375)

A resistance to a diffusing innovation, which can be of a physical, cultural, or psychological nature.

Geometric boundary (378)

A border that follows a straight line, usually a parallel of latitude or a meridian of longitude.

Physiographic boundary (378)

A border coinciding with a physical feature on the landscape, such as a river, coastline, or mountain ridge.

OPEC (379-380)

The cartel called the Organization of Petroleum Exporting Countries, whose 13 member-states are mapped in Fig. 6-13.

Basin irrigation (382)

An ancient irrigation method of the lower Nile Valley involving the trapping and later release of floodwaters.

Perennial irrigation (382-383)

The more modern Egyptian irrigation technique using dams and levees to store and use floodwater throughout the year.

Fellaheen (385)

Egypt's peasant farmers.

Remote sensing (386-387 box)

The technique used to gather information about the earth's surface by using satellite-based instruments; since each surface feature and land-use activity emits its own spectral signature, highly detailed data can be collected by remote sensors.

Maghreb (388)

The western region of this realm comprising the Northwest African countries of Morocco, Algeria, and Tunisia.

Tell (390)

The lower slopes and coastal plains of Northwest Africa between the Atlas Mountains and the sea.

Kibbutz (393)

An Israeli collectivized farm settlement.

Nomadism (398 box)

A group of people who move cyclically among a set of places, usually practicing pastoralism.

Cycle theory (400-401)

Samuel van Valkenburg's theory of state evolution entailing four stages: organization, expansion, stabilization, and disintegration.

Qanat (404)

Underground water tunnels leading from mountains to nearby dry flatlands.

Ecological trilogy (406)

Paul Ward English's generalization of the linkages of traditional Iranian society, involving interactions among urbanites, villagers, and nomads.

Sahel (407-408)

Arabic for "border," refers to the heart of the African Transition Zone--an east-west belt stretching across the southern edge of the Sahara Desert.

SELF-TESTING QUESTIONS

Cover this side of the page with a sheet of paper. Uncover each line after you have attempted to answer the question in the left column. If necessary, refer to the textbook page(s) listed at the right.

Question	Answer	Page
REALM CHARACTERISTICS	REALM CHARACTERISTICS	
Why is "Dry World" a misleading title for this realm?	Most of its people concentrate in the few well-watered areas.	361
Why is "Arab World" equally misleading?	This is a linguistic term which does not fit much of the region.	361
Why is "World of Islam" just as unsatisfactory?	Many exceptions exist to the Moslem religion, and major internal divisions among sects exist.	363-64
Why is "Middle East" not much of an improvement?	It is imprecise, and reflects a European perceptual bias.	364

Question	Answer	Page
CULTURAL GEOGRAPHY	CULTURAL GEOGRAPHY	
What are the five major themes of cultural geogaphy?	Culture; culture areas; cultural landscapes; culture history; and cultural ecology.	362
What is the central concern of cultural landscape interpretation?	The physical imprint of material culture on the land, which forms the geographic content of a culture region.	362
How do popular and folk culture differ?	The former is the ever-changing mass culture of advanced societies; the latter is the much more strongly traditional way of life found in isolated rural areas.	362
What is possibilism?	The belief that human cultures are free to choose from several "possibilities" offered by the natural environment.	363
HISTORICAL GEOGRAPHY	HISTORICAL GEOGRAPHY	
What were the three culture hearths of North Africa/Southwest Asia?	The lower Nile Valley, the Tigris-Euphrates Lowland (Mesopotamia), and the lower Indus Valley.	365
Name some of the major agricultural innovations that originated in these hearths.	Wheat, rye, barley, peas, beans, grapes, olives, apples, peaches, horses, pigs, sheep, and goats.	367
When did the Prophet Mohammed live?	571 - 632 A.D.; he became a religious leader in 613 A.D.	369
What are the "five pillars" of Islam?	Repeated statement of the creed; frequent prayer; a month of daytime fasting, alms-giving, and a pilgrimage to Mecca.	369
What are the leading centers of Judaism and Christianity in this realm?	Jews are clustered in Israel; Christians dominate portions of Lebanon and Ethiopia.	371
SPATIAL DIFFUSION	SPATIAL DIFFUSION	
What is an innovation-wave?	The stages of Hagerstrand's model, showing the progress of a diffusing phenomenon, which pulses outward from its hearth across the adopting region.	372
What are the two major types of spatial diffusion?	Relocation diffusion (diffusion by migration) and expansion diffusion (diffusion within a fixed population).	372

What are the two forms of expansion diffusion?	Contagious diffusion (distance-controlled local diffusion) and hierarchical diffusion (coursing downward within a national urban hierarchy).	372-73
What are the three categories of barriers to diffusion?	Physical, cultural, and psychological.	375
MIDDLE EAST OIL	MIDDLE EAST OIL	
How much of the world's oil reserves are contained within this realm?	Still over 50 percent.	378
Name the thirteen members of OPEC as of the end of 1987.	Algeria, Ecuador, Gabon, Indonesia, Iran, Iraq, Kuwait, Libya, Nigeria, Qatar, Saudi Arabia, United Arab Emirates, and Venezuela.	378-79 map
What is the richest oil state?	The United Arab Emirates, whose 1983 per capita income was $28,110; neighboring Qatar is at the same level.	380
Which major country was suspended from the Arab League in 1979?	Egypt, because of its peace treaty with Israel; League headquarters were also moved from Cairo to Tunis.	381
EGYPT AND NEIGHBORS	EGYPT AND NEIGHBORS	
How important is the Nile to this country?	This river is Egypt's lifeline; valley and delta are home to well over 90 percent of the population.	382
How do basin and perennial irrigation differ?	The former is an ancient method of trapping floodwaters; the latter is a modern system based on dams, of which the Aswan Dam is most prominent.	382-83
Why is Egypt so prominent an Arab country?	Its pivotal location, maintained throughout history; the Nile also fostered the growth of the Middle East's largest national population.	385
Name Egypt's six major subregions.	Nile Delta; Middle Egypt (including Cairo); Upper Egypt; Western Desert; Eastern Desert/Red Sea Coast; Sinai Peninsula.	385-86
THE MAGHREB	THE MAGHREB	
What does this term mean? Name its constituent countries.	Literally "western isle," based on the Atlas Mountains standing above the flat Sahara; Morocco, Algeria, and Tunisia.	388

Question	Answer	Page
Who are the Berbers? Who were the Moors?	The Maghreb's oldest inhabitants, who adapted to Islam; the Berber-Arab alliance (Moors) pushed into Iberia, but eventually were driven out by the Europeans.	389
What are <u>bidonvilles</u>?	The poverty-stricken shantytowns that ring the region's cities.	390
THE MIDDLE EAST	THE MIDDLE EAST	
Name the 5 countries of this region.	Israel, Syria, Lebanon, Jordan, and Iraq.	391
How old is Israel as a modern state?	It was founded amid much turmoil in 1948, and has lived with threatening neighbors ever since.	391
What is the distribution of Palestinians in the Middle East?	In the mid 80s, well over 3 million are dispersed throughout the region; Jordan contains 800,000, and Syria, Kuwait, and Lebanon contain large concentrations; even Israel itself is home to 560,000.	394 box
How has Israel's development fared vis-a-vis its neighbors?	Far better; while the Arab countries remain relatively unproductive and mired in poverty, Israeli prosperity and economic growth is impressive.	393
What are Israel's industrial specialties?	Despite an inferior resource base, Israel (like Japan) thrives; highly-skilled labor is the specialty for industries such as diamond cutting.	393
Why is Lebanon so beset by internal strife?	The old political accommodation between the Moslems and Christians no longer fits a new reality in which Moslem groups (themselves disunified) have come to dominate the population.	394-95
What are the leading problems of Syria, Jordan, and Iraq?	Syria's productive areas are few and far between, and its army is bogged down in Lebanon; Jordan is beset by a population in which refugees outnumber natives, and survives only with foreign aid; government corruption, land-use inequities, and a ruinous war with Iran (still continuing in 1988).	396-97

Question	Answer	Page
ARABIAN PENINSULA	ARABIAN PENINSULA	
Name the countries of this region.	Saudi Arabia, Kuwait, Bahrain, Qatar, United Arab Emirates, Oman, and the two Yemens.	397-98
What are some major recent Saudi development projects?	The construction of an ultramodern circulatory system; the growth of the petrochemical industry; the new cities of Jubail and Yanbu.	400
NON-ARAB NORTH	NON-ARAB NORTH	
Name the states of this region.	Turkey, Iran, Cyprus, and Afghanistan.	401
Why has Turkey remained aloof from other Arab countries?	Greater European influences shaped culture here; Ataturk's revolution stressed the detachment from other Islamic states, and that tradition endures.	401
Who are the contestants in the Cyprus struggle?	Greeks who seek union with Greece; Turks who want ties to Turkey.	405 box
How do Iran's Moslems differ from those of other Arab countries?	Most belong to the Shi'ite sect, comprising about 10 percent of all Moslems; elsewhere the majority Sunni sect is dominant.	403
How has modernization affected Iran?	Surprisingly little; it was also a contributory factor to the revolution that deposed the Shah.	406
What is the <u>ecological trilogy</u>?	The interactions among Iran's urban, village, and nomadic communities.	406
Which states established the buffer role for Afganistan?	19th century Britain and Russia, for which the country was a neutral zone between their Asian spheres of influence.	407
AFRICAN TRANSITION ZONE	AFRICAN TRANSITION ZONE	
What is the extent of this region?	The east-west band of states along the southern margin of the Sahara, from Mauritania and Senegal in the west to Ethiopia and Somalia in the eastern "African Horn."	407
Which economic activity is dominant?	Cattle herding, a livelihood always vulnerable to drought and famine in the steppes of the Sahel.	407-8
Give an example of irredentism in this region.	Pan-Somaliism, which has spawned warfare and misery for millions of Somalis in the contested Ogaden region of Southern Ethiopia.	408-9

MAP EXERCISES

MAP COMPARISON

1. Compare the map of world religions (Fig. 6-3, pp. 366-367) to the map of geographic realms (pp. 42-43). Which realms are clearly identifiable with a major religion? Which realms exhibit multiple religions that portend internal cultural conflicts? Double-check your observations against the world population maps (pp. 26-29), underscoring those realms associated with large population clusters.

2. Carefully compare the two diffusion patterns in Fig. 6-11 (pp. 374-375) by preparing a table showing cholera outbreak by city in chronological order. Besides the main patterns discussed in the text of the box, do you detect any additional aspects of contagious and/or hierarchical diffusion in your table?

3. Prepare a brief essay on the current distribution of world coal, petroleum, and natural gas reserves, based on your analysis of Fig. 6-13 (pp. 378-379). How does this compare to patterns of production and consumption as shown in the green-and-white bar graphs on p. 379.

4. Review the map-reading box on pp. 51-53, and then reread the box on remote sensing on pp. 386-387. How does a remotely-sensed image differ from a map? In what ways might a map have been better for showing the Nile Delta on p. 386? What are the advantages of this satellite view?

5. Compare the Addis Ababa map on p. 409 with the Von Thunen maps in the box on pp. 76-78. How does the Ethiopian distribution differ from the ideal model layout, and what similarities and differences do you observe between the Addis Ababa example and the modern European case?

MAP CONSTRUCTION

(Use outline maps at the end of this chapter)

1. In order to familiarize yourself with North African/Southwest Asian physical geography, place the following on the first outline map:

    a. Rivers: Nile, White Nile, Blue Nile, Tigris, Euphrates, Orontes, Jordan, Shatt al Arab

    b. Water bodies: Red Sea, Persian Gulf, Suez Canal, Gulf of Aqaba, Gulf of Suez, Bab el Mandeb Straits, Strait of Hormuz, Strait of Gibraltar, Gulf of Sidra, Caspian Sea, Black Sea, Mediterranean Sea, Lake Nasser, Lake Tana, Sudd, Dead Sea, Bosporus Strait, Gulf of Aden, Gulf of Oman, Arabian Sea

    c. Land bodies: Sinai Peninsula, Cyprus, Horn of Africa, Socotra, Bahrain, Canary Islands

d. Mountains and Deserts: Atlas Mountains, Tibesti Mountains, Libyan Desert, Sahara Desert, Arabian Desert, Rub al Khali (Empty Quarter), Anatolian Plateau, Ahaggar Mountains, Ethiopian Highlands, Hejaz Mountains, Negev Desert, Plateau of Iran, Elburz Mountains, Zagros Mountains, Pamir Mountains, Hindu Kush Mountains, Nubian Desert, An Nafud Desert

2. On the second map political-cultural information should be entered as follows:

   a. Label each country of the realm, and label and locate each capital city with the symbol *

   b. Color each country according to its major religion, using information and appropriate color symbols from the map on text pp. 366-367 (be sure to differentiate between the Sunni and Shiah sects in Moslem states)

3. On the third outline map, economic-urban information should be entered as follows:

   a. Cities (locate and label with the symbol • /capitals should be shown with symbol * again): Casablanca, Marrakech, Rabat, Tangiers, Oran, Algiers, Tunis, Tripoli, Benghazi, Ndjamena, Niamey, Bamako, Timbuktu (Tombouctou), Nouakchott, Port Sudan, Khartoum, Addis Ababa, Asmera, Djibouti, Mogadishu, Cairo, Alexandria, Port Said, Ismailia, Aswan, Riyadh, Mecca, Medina, Dhahran, Jubail, Yanbu, Kuwait, Manama, Doha, Abu Dhabi, Muscat, Aden, San'a, Amman, Jerusalem, Haifa, Tel Avia-Jaffa, Eilat, Aqaba, Beirut, Sidon, Damascus, Aleppo, Nicosia, Baghdad, Basra, Kirkuk, Istanbul, Ankara, Izmir, Adana, Zonguldak, Tehran, Tabriz, Abadan, Mashhad, Shiraz, Kabul, Herat

   b. Economic regions (identify with circled letter):

      A - Eritrea
      B - Neutral Zone
      C - West Bank
      D - Mesopotamia
      E - Lower Egypt
      F - Gaza
      G - Wakhan Corridor
      H - Asia Minor
      I - Gezira
      J - "The Tell"
      K - Persian Gulf oilfield
      L - Jezira

PRACTICE EXAMINATION

SHORT-ANSWER QUESTIONS

Multiple Choice

1. Which of the following countries is part of the Maghreb region:

   a) Libya
   b) Egypt
   c) Algeria
   d) Turkey
   e) Gibraltar

2. Which of the following rivers flows through Iraq?

   a) Euphrates
   b) Jordan
   c) Nile
   d) Indus
   e) Bosporus

3. Kemal Ataturk is most closely identified with the city of:

   a) Cairo
   b) Mecca
   c) Tehran
   d) Ankara
   e) Baghdad

True-False

1. The Ogaden region is hotly contested by Israel and Syria in the mid-1980s.

2. The region known as the Sahel is most correctly located in the African Transition Zone.

3. Hierarchical diffusion involves the spreading of an innovation across a large area, occurring within a country's overall urban system.

Fill Ins

1. The eastern Mediterranean island of ______________ is hotly contested by Greece and Turkey.

2. ______________ diffusion depends upon a migrating population to transmit an innovation.

3. The Moslem sect that dominates life in Iran today is the ______________.

Matching Question on the Realm's Countries

_____	1.	Home of the fellaheen	A.	Israel
_____	2.	Maghreb kingdom	B.	Syria
_____	3.	North Africa's most radical state	C.	Somalia
_____	4.	Headquarters of Ottoman Empire	D.	Turkey
_____	5.	White and Blue Nile convergence	E.	Libya
_____	6.	Christian-Moslem turmoil	F.	Saudi Arabia
_____	7.	Jubail project	G.	Afghanistan
_____	8.	Led by Bourguiba (through 1987)	H.	Algeria
_____	9.	Irredentism directed against Ethiopia	I.	Iraq
_____	10.	Stronghold of Kurds	J.	Jordan
_____	11.	Zionist movement	K.	Egypt
_____	12.	Tell Atlas	L.	Morocco
_____	13.	Former Russian-British buffer	M.	Lebanon
_____	14.	Refugees outnumber natives	N.	Tunisia
_____	15.	Lost Golan Heights to Israel	O.	Sudan

ESSAY QUESTIONS

1. Geographical definitions of this realm have been difficult to arrive at and are still not entirely satisfactory. Compare and contrast the "Dry World," "Arab World," and "World of Islam" definitions, highlighting the strengths and weaknesses of each. Does the appellation "Middle East" add anything useful to the argument, and why is it not used in the text as the realm's name?

2. The influence of the Moslem faith has been a dominant shaping force in the history of the realm for the past 1400 years. Elaborate on the diffusion and acceptance of Islam in North Africa/Southwest Asia since 600 A.D., highlighting cultural-geographical similarities and differences.

3. Define and discuss the two major types of expansion diffusion, citing examples of each. What do the two maps of cholera diffusion in the United States tell us about national communication channels in 1832 and 1866?

4. How has the production of oil affected the differential economic development of the realm's countries?

5. Discuss the regional problems of the Maghreb, highlighting environmental variables, internal political differences, resource distributions, and economic development opportunities.

6. Discuss the changing political geography of Israel since its creation in 1948. How has this country strengthened itself since the 1967 War, and what is the present likelihood of a more stable coexistence with its Arab neighbors?

7. Compare and contrast the political and economic experiences of Saudi Arabia and Iran in the past ten years, and evaluate the chances for meaningful modernization and development in the two countries during the coming few years.

8. Review the problems of the African Transition Zone and the chances for their resolution in the immediate future.

TERM PAPER POINTERS

The "Term Paper Pointers" section of Chapter 1 in this Study Guide offered suggestions about approaching research and writing on geographic realms and their components, and you may wish to consult this material if you are undertaking a report on a North African or Southwest Asian region.

The best approach to this sprawling realm is again by consulting a basic regional geography. The literature, though dated for so politically dynamic a part of the world, is more than adequate though somewhat dominated by a British point of view--Fisher, Beaumont et al., Longrigg, and Cressey. Atlases worth looking at are those by Adler et al., al Faruqi [both sources], and Gilbert; the two bibliographies by Miller & Miller will also be helpful. Individual regions are treated by Clarke & Bowen-Jones, Cloudsley-Thompson, Cottrell, Parker, and Thompson & Adloff. Single countries are examined by Al-Farsy, Allan, Amirahmadi & Parvin, Ansari, Deeb & Deeb, Devlin, Girardet, Gordon, Haddad, Hammond, G. Harris L. Harris, Helms, Hiro, Mackey, Mottahedeh, Orni & Efrat, Peck, Perlmutter, and Wubneh & Abate. Additional perspectives on the realm can be obtained from Asad & Owen (overview), Coon (anthropology), Ahmed, Rahman, and Wright (Islamic traditions); two penetrating journalistic efforts are Lamb and Shipler [1986].

Numerous topical treatments of the realm's geography are available as well. Economic geography and development is found in Beaumont & McLachlan, El Mallakh, Fahim, and Horvath. The geography of energy, an important new subdivision of economic geography with obvious significance for understanding this realm, is handled in Calzonetti & Solomon, Cook, Smil, and Ruedisili & Firebaugh. Urban geography is elaborated by Abu-Lughod, Blake & Lawless, Efrat, Karan & Bladen, and Saqqaf. Remote sensing is covered by Holz and Jensen. The geography of avid lands is treated by Cloudsley-Thompson, Glantz, and Heathcote. The Arab-Israeli conflict is covered in Adler, S. Cohen, Gilbert, Laqueur & Rubin, Perlmutter, Peters, and Shipler [1986]. Politico-geographical issues are surveyed in S. Cohen, Cowell, Ghanem, Drysdale & Blake, Naff & Matson, and Prescott. Historical geography is treated in Isaac, Murphey, Sauer, and both books by Wagstaff. The current dilemmas of famine and refugees are covered in Currey & Hugo and Rogge.

The cultural geography Systematic Essay is also a fruitful basis for a term paper. The key references and source works are Dohrs & Sommers, Mikesell, Spencer, and Wagner & Mikesell. Specialized topics are treated by English, Friedlander & Goldscheider, Weekes, and Simoons. The closely-related subject of spatial diffusion--the focus of the following Special Exercise--is covered in Gould, Brown, Rogers, Chapman, and Hagerstrand; specific applications can be found in Y. Cohen, Hunter & Young, and both items by Pyle.

SPECIAL EXERCISE

The purpose of this exercise is to have you trace the spatial diffusion of a specific phenomenon. You will need to review the basic principles of diffusion, decide which phenomenon to study, collect relevant data on the arrival of the phenomenon at specific locations, and map the overall diffusion pattern. It is also recommended that you consult your instructor, who can offer valuable suggestions about the tasks involved in this project.

The best place to review the basic principles of spatial diffusion is Peter Gould's Spatial Diffusion (ref. on p. 412). This brief overview is still in print from the Association of American Geographers, 1710 16th Street, N.W., Washington, D.C. 20009; it is also available as Chapter 11 in a textbook entitled Spatial Organization by Ronald Abler, John Adams, and Peter Gould (Prentice-Hall, 1971)--most libraries will contain copies. Gould's monograph contains numerous examples in its second half, and this section of the work will provide many ideas about approaching the subject. It is also strongly recommended that you read an article in the literature on the diffusion of a phenomenon: Pyle and Hunter & Young (refs. p. 413) are excellent for this purpose, and your instructor can recommend others.

With theory and method covered, you are now ready to begin searching for the empirical information you will need. The articles you have been reading provide clues as to where data can be obtained; if you cannot find data for every year, it is perfectly all right to use data for a regular interval of say every five years. For example, if you wanted to study the diffusion of FM/stereo radio stations in the United States, directories would be consulted for the appropriate five-year intervals. Directories are often the most valuable sources for information on the spreading of a phenomenon--many companies such as the motel chains publish annual listings which are helpful in collecting diffusion data when they are examined in chronological order. Let your imagination be your guide--you know what kind of information you are looking for, and the materials you have read provide a number of ideas about sources for relevant data.

Once you have collected sufficient data, you should map them using a cartographic technique illustrated in Model Box 4 (text pp. 372-375). Map needs will hopefully be satisfied by photocopying one or more of the outline maps in this Study Guide (if larger-scale maps are needed, your instructor can help you locate them). When your map is finished it is also a good idea to study its patterns carefully and briefly write down your observations. Some questions worth considering are: Is a clear diffusion pattern discernible? How well does this pattern concur with the expectations drawn from the theoretical principles of spatial diffusion? Is hierarchical or contagious diffusion the most important process you observed, and why? What contribution to our knowledge of diffusion processes does your study make?

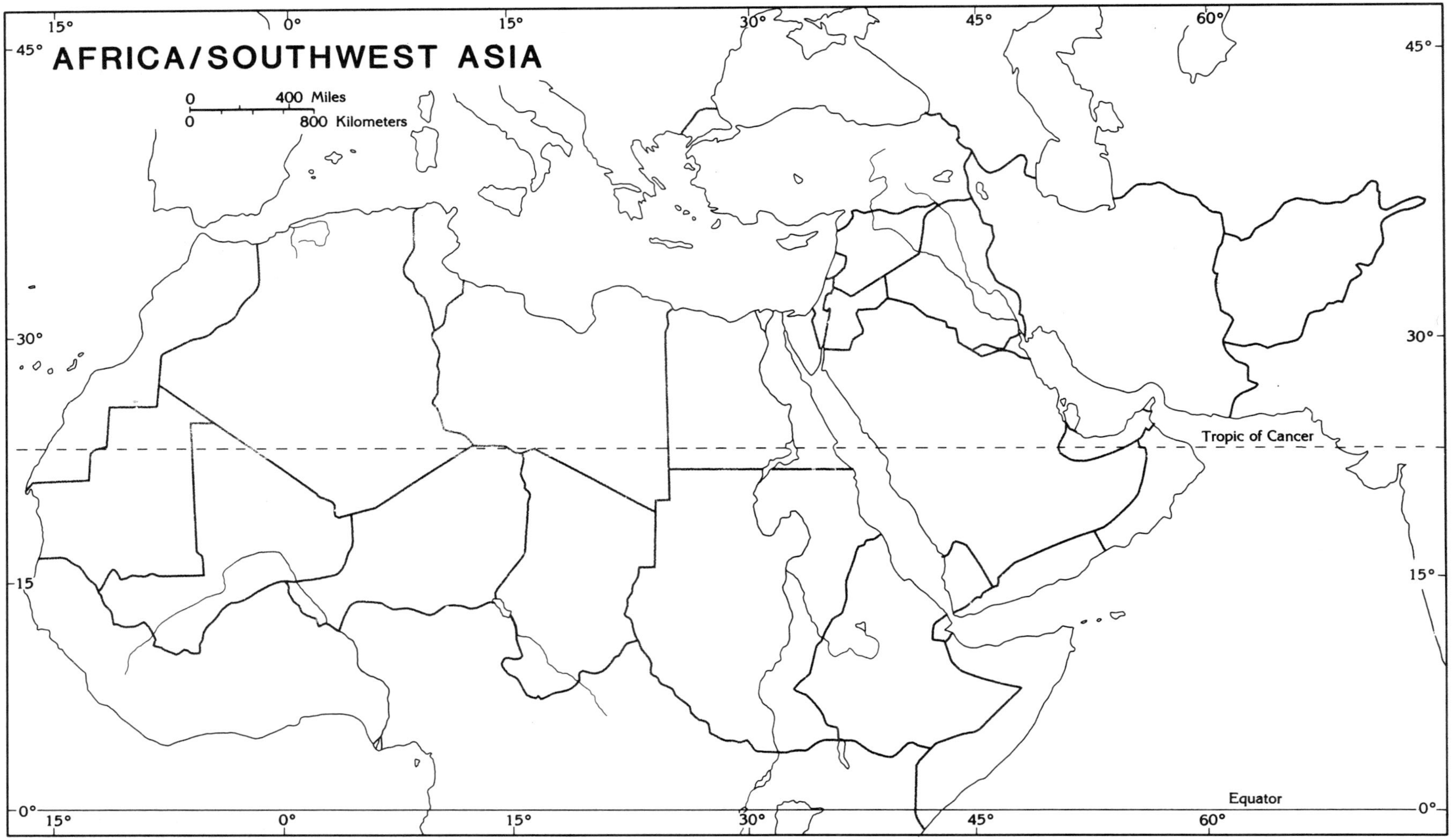
AFRICA/SOUTHWEST ASIA
0 400 Miles
0 800 Kilometers
Tropic of Cancer
Equator
15°
0°
15°
30°
45°
60°
45°
30°
15°
0°

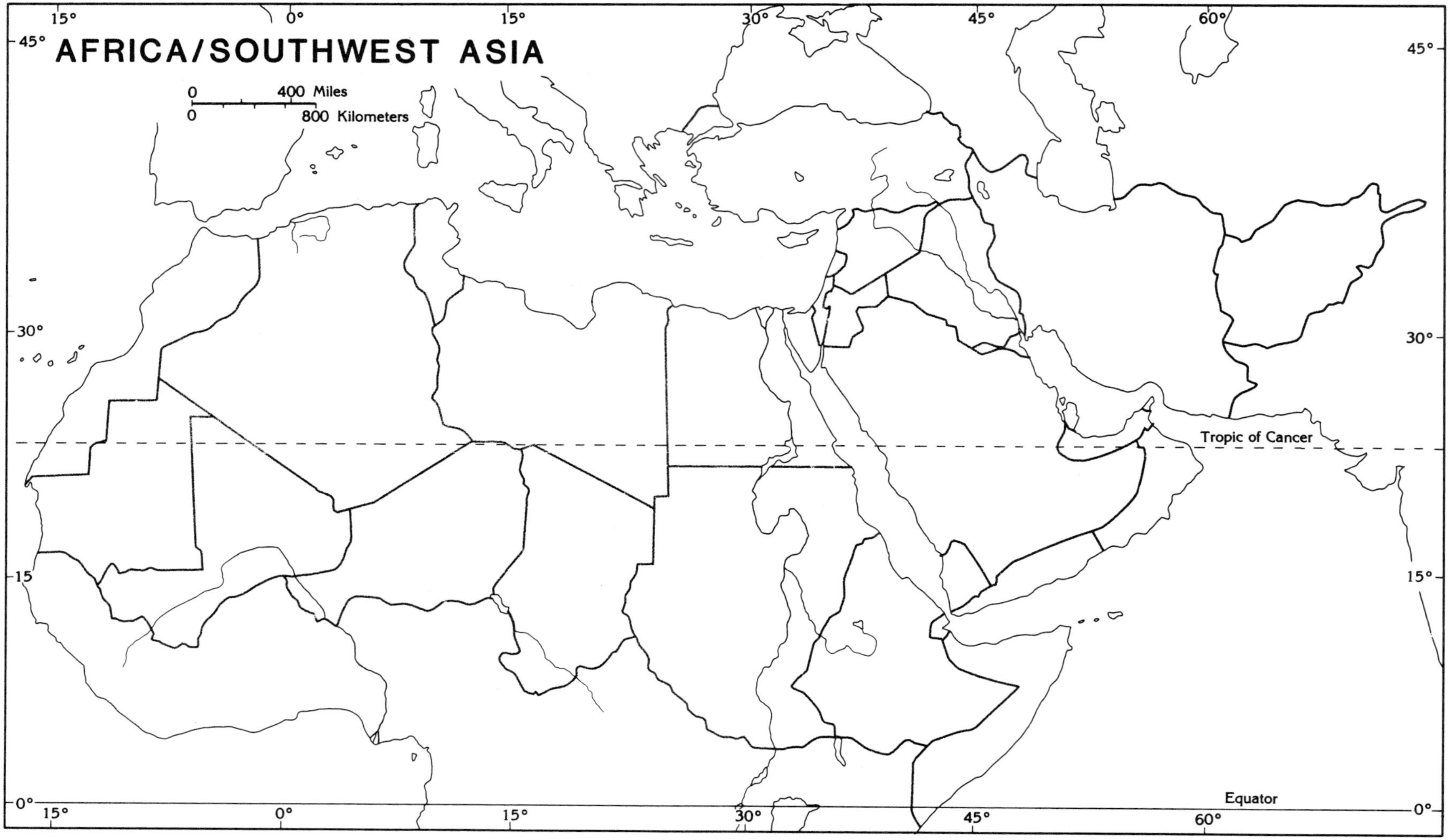
AFRICA/SOUTHWEST ASIA
0 400 Miles
0 800 Kilometers
Tropic of Cancer
Equator
15°
0°
15°
30°
45°
60°
45°
30°
15°
0°

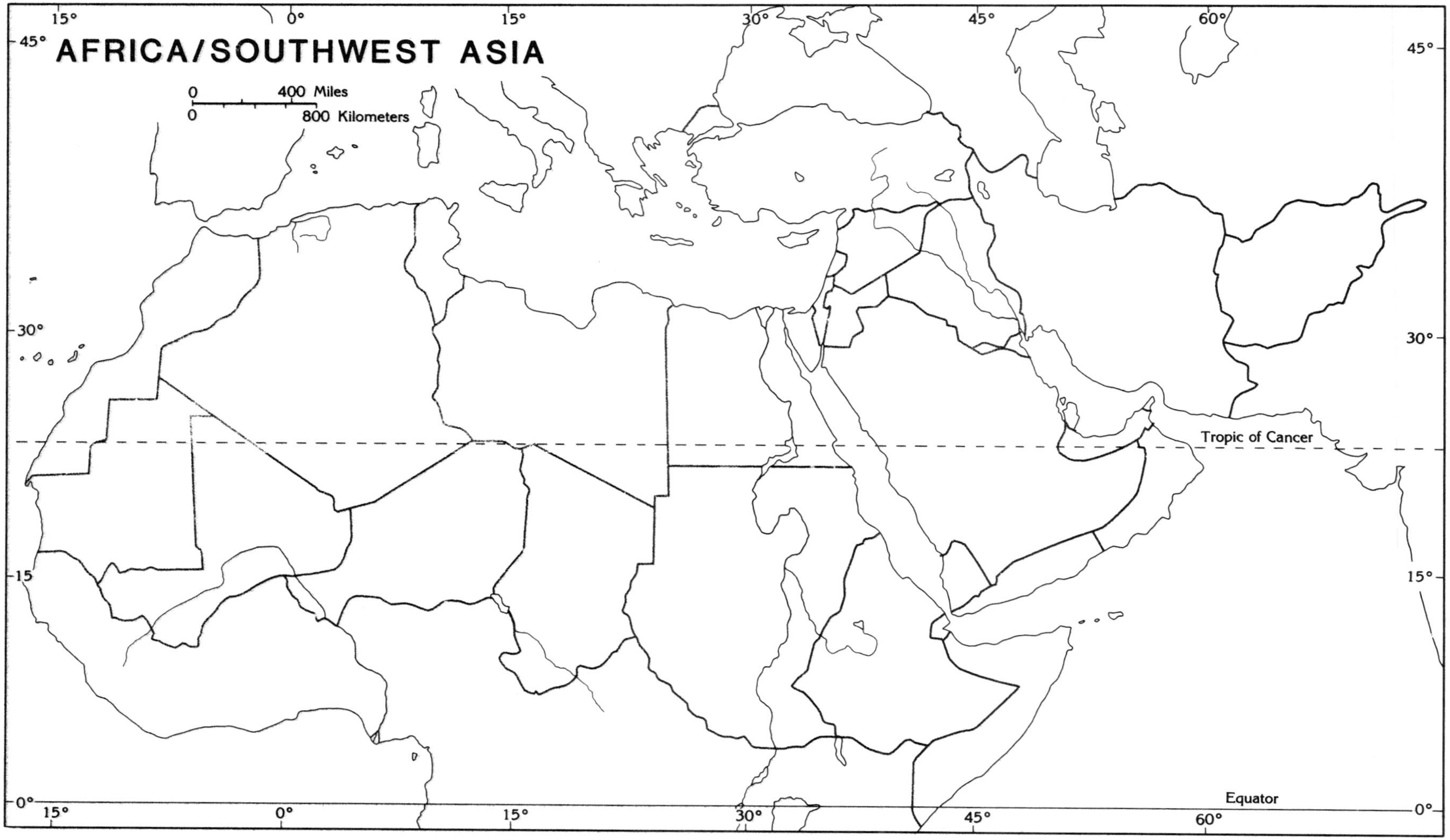
AFRICA/SOUTHWEST ASIA
0 400 Miles
0 800 Kilometers
Tropic of Cancer
Equator
15°
0°
15°
30°
45°
60°
45°
30°
15°
0°

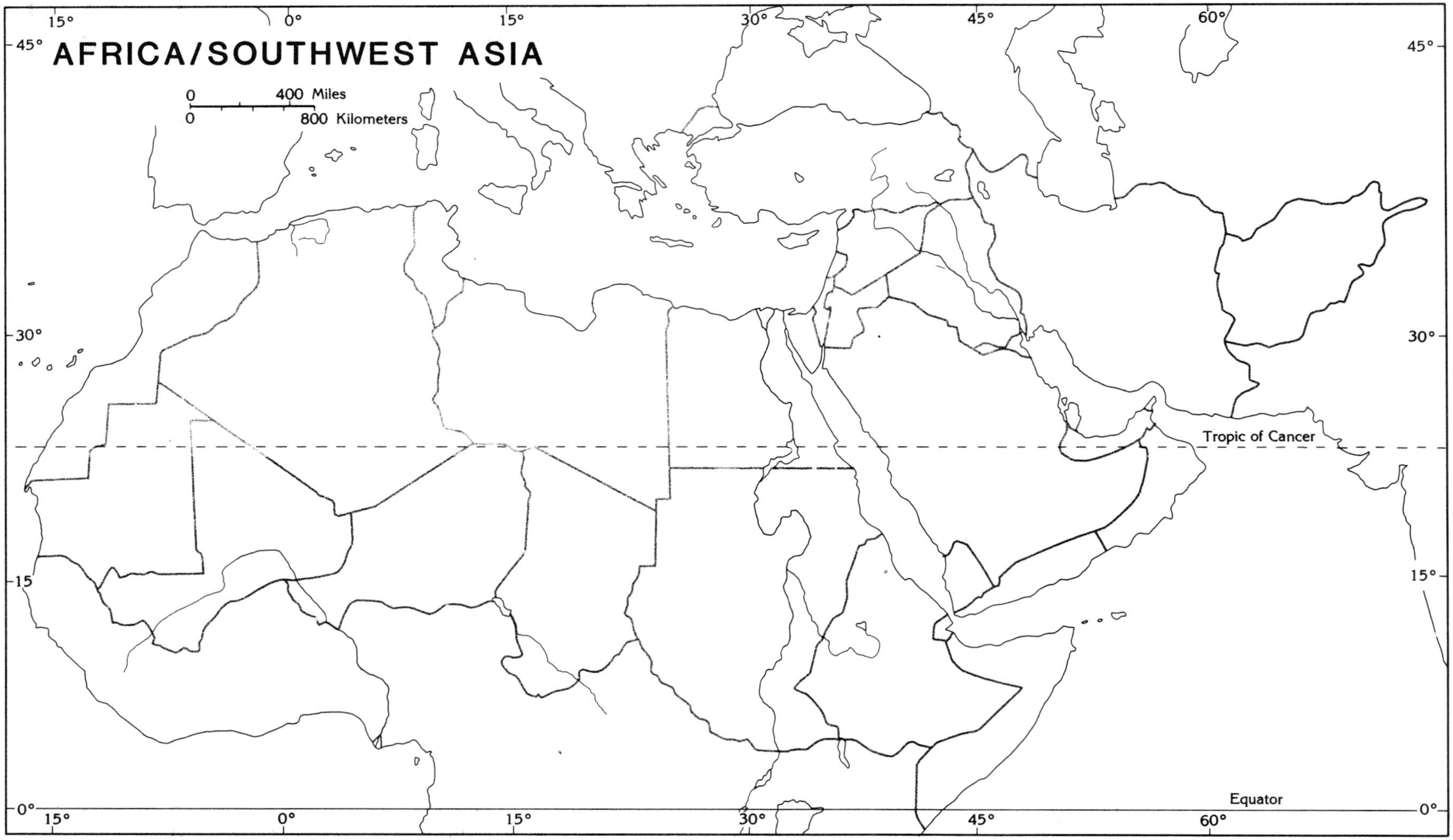
AFRICA/SOUTHWEST ASIA
0 400 Miles
0 800 Kilometers
15°
0°
15°
30°
45°
60°
45°
30°
15
0°
Tropic of Cancer
Equator

CHAPTER 7: SUBSAHARAN AFRICA — ADVERSITIES AND ADVERSARIES

OBJECTIVES OF THIS CHAPTER

Chapter 7 treats Subsaharan Africa, a problem-ridden realm in which life for the masses is almost always difficult. The widespread occurrence of chronic illnesses here makes the Systematic Essay on medical geography particularly appropriate. Following the introduction, Africa's environmental base is covered and highlights physiography, disease patterns, and the agricultural potential as well as risks. Historical geography comes next, examining pre-European cultures, the colonial transformation after 1450, and the independence era which has endured for the past three decades. The regional survey covers the rest of the chapter, focusing in turn on West, East, Equatorial, and Southern Africa. The Republic of South Africa is treated separately in a Vignette that immediately follows this chapter (both here and in the text).

Having learned the regional geography of Subsaharan Africa, you should be able to:

1. Appreciate the many concerns of contemporary medical geography and their applications to Africa.

2. Understand the interpretation of African physiography and hydrography that is offered by the continental drift hypothesis.

3. Understand the occurrences and distributions of Africa's endemic diseases and their impacts on daily life.

4. Grasp the importance of agriculture in African economic life, and the many environmental risks that farming and herding are exposed to.

5. Understand the course of Subsaharan African history from the indigenous cultures through the colonial and contemporary eras.

6. Explain the political territorialization of modern Africa as a legacy from the colonial era that has just ended.

7. Understand the cultural and economic trends that have shaped the regional structuring of West, East, Equatorial, and Southern Africa.

8. Locate the leading physical, cultural, and economic-spatial features of the realm on an outline map.

GLOSSARY

Medical geography (416)

The subdiscipline concerned with the spatial aspects of health and illness.

Disease ecology (416)

Concerned with the study of the manner and consequences of interaction between the environment and the causes of morbidity and mortality.

Morbidity (416)

The condition of illness.

Mortality (416)

The occurrence of death.

Language family (418)

A linguistic classification term: languages believed to have a shared origin are grouped into a family (even closer relationships put them into subfamilies).

Race (419)

A non-cultural term, referring to the physical features of a breeding population group.

Rift valley (421)

Geologic trenches formed when huge parallel cracks or faults occur in the earth's crust, causing in-between strips of land to sink to form extensive linear valleys.

Great Escarpment (424)

The edge of the African plateau that forms a steep cliff, hundreds of miles in length, stretching north from southern South Africa to Tanzania on the east coast and as far as Zaire on the Atlantic coast.

Continental drift (425)

The hypothesis underlying the formation of Africa's physiography and hydrography. The break-up of the super continent called Gondwana and the subsequent drifting apart of its landmass components; Africa occupied the heart of Gondwana, and its physical geography is still shaped by this geologic functioning as the core-shield.

Pandemic disease (427)

A world-scale disease, orginating as an outbreak in a locality but spreading contagiously to eventually blanket the globe (such as influenza).

Endemic disease (427)

A chronic illness existing in equilibrium with a host population, such as malaria.

Dhows (433)

Wooden boats with triangular sails, plying the seas of the Indian Ocean between the Arabian and the East African coasts.

Berlin Conference (438)

The 1884 international conclave, convened by Germany's Otto von Bismarck, which superimposed on Africa a system of political boundaries agreed to by all colonial powers; amazingly, the scheme has survived the past century and this political fragmentation underlies the realm's contemporary territorial structuring.

Condominium (440)

The shared administration of a territory by two or more governments.

Negritude (440)

Philosophy promoting black cultural pride, with roots in Francophone West Africa and the Caribbean.

Periodic market (448-449)

Village market that opens every few days; part of a regional network of similar markets in rural, preindustrial societies where goods are brought to market on foot and barter remains a major mode of exchange.

Sequent occupance (452 box)

The idea that successive societies leave their imprints on a place, each contributing to the cumulative cultural landscape.

SELF-TESTING QUESTIONS

Cover this side of the page with a sheet of paper. Uncover each line after you have attempted to answer the question in the left column. If necessary, refer to textbook page(s) listed at the right.

Question	Answer	Page
MEDICAL GEOGRAPHY	MEDICAL GEOGRAPHY	
What is disease ecology?	The study of interactions between the environment and the causes of illness and death.	416
What are disease agents and vectors?	Disease agents are the pathogens themselves; vectors are the intermediate transmitters of pathogens.	416
How can contagious diffusion be a useful concept for combatting disease?	Many diseases such as cholera and meningitis spread in such a manner; knowing likely diffusion routes allows health planners to target preventive-measure areas.	416

AFRICAN ENVIRONMENTS	AFRICAN ENVIRONMENTS	
What is a rift valley?	A trench formed when cracks or faults occur in the earth's crust, and in-between strips sink down to create linear valleys.	421
What is the Great Escarpment?	The margin of Africa's plateau, especially in Southern Africa, where the land surface drops sharply to the coastal lowland.	424
What is continental drift? How does it help to understand Africa's physiography?	The hypothesis that all continents were once joined into a single supercontinent (Pangaea); Africa was located centrally in the southern component (Gondwana); the giant landmass broke apart at least 100 million years ago, with its continental components "drifting" away toward their present locations. Africa's unusual topographic and drainage patterns are best explained by this hypothesis.	425
Describe Africa's broad patterns of climate and vegetation.	They are symmetrically distributed about the equator, which bisects the continent. The humid tropics are largely restricted to the western half since East Africa is largely an elevated plateau; as one moves poleward from the tropics, steppes, significant deserts, and narrow strips of Mediterranean climates prevail.	427
What is the difference between an epidemic and a pandemic disease?	The former is a local or regional phenomenon; an outbreak of near-global proportions is described as pandemic.	427
What is Africa's worst endemic disease?	Malaria, whose vector (carrier) is a mosquito that prevails in most of Africa's inhabited areas.	427
How is sleeping sickness transmitted?	By the tsetse fly, native to most of tropical Africa, which infects both humans and livestock (see Fig. 7-11).	428
Where is river blindness a major hazard?	In the savanna belt south of the Sahara, particularly in West Africa; it is spread by flies carrying parasitic worms.	428
AFRICAN AGRICULTURE	**AFRICAN AGRICULTURE**	
Why is water supply such a problem in this tropical, island continent?	The rainforests are not good environments for farming, and evapotranspiration is very high in the wet tropics; elsewhere, drought is a constant threat.	430

Why is African farming generally unproductive?	It is almost exclusively of the traditional subsistence variety, heavily localized, and tied to an age-old, inefficient technology.	431
What is the role of cattle in much of Africa?	Less important as a food source than as a measure of prestige of their owners in community life; quantity of animals is more important than their health and quality.	431
What is the prospect for agricultural improvement?	Slight in the foreseeable future; cash cropping has increased slightly as urbanization has intensified, but the life of the farmer remains difficult.	432
HISTORICAL GEOGRAPHY	HISTORICAL GEOGRAPHY	
Why is so little known about Subsaharan Africa before A.D. 1500?	The absence of written histories is the main cause, exacerbated by neglect and destruction of traditions and artifacts following the onset of colonialism and the disruptions of the slave trade.	433
Does this mean that pre-colonial Africa had no organized cultures and traditions?	Definitely not; on the contrary, sophisticated artifacts throughout Africa suggest advanced societies, widespread trade, and rich cultural legacies. West Africa, especially, was the scene of impressive cities, trade networks, and states of amazing durability such as old Ghana and Songhai.	433-34
Where did the first European contacts occur?	After 1480, Europe's ships began to ply the waters of the entire west coast of Africa as the route to southern Asia opened. Way-station ports opened, and soon African middlemen began to organize trading of slaves, minerals, ivory, and spices.	434-35
How did the subsequent European penetration proceed?	Slowly at first, but after 1800 European powers finally laid claim to all of Africa's territory; expansion of spheres of influence soon caused disagreements but the 1884 Berlin Conference solved things peaceably (see p. 438).	436-37
What administrative systems did the Europeans employ in colonial Africa?	Indirect rule of colonies, protectorates, and territories (Britain); paternalism (Belgium); centralization and assimilation toward the mother country (France); rigid economic control (Portugal).	438-44

WEST AFRICA	WEST AFRICA	
Which countries are included in this region?	Mauritania, Mali, Niger, Senegal, Gambia, Guinea-Bissau, Guinea, Sierra Leone, Liberia, Burkina Faso, Ivory Coast, Ghana, Togo, Benin, and Nigeria.	445
What regional unities exist?	A remarkable cultural-historical momentum; parallel east-west environmental belts, with much interaction perpendicular to them; an early economic orientation to world commodity markets.	446
Where are West Africa's main population concentrations?	The coasts of the African "bulge," and certain drier interior zones such as northern Ghana and Nigeria's Moslem north.	447
Who are Nigeria's three leading peoples?	The Yoruba of the southwest, the Ibos of the east, and the Hausa-Fulani of the semiarid north.	447
What are some recent signs of Nigerian progress following the Biafran disaster of the 1960s?	Greater political stability of its revised federal system; the construction of a new centrally located, compromise capital; exploitation of rich oil reserves; the improvement of its circulatory system.	447-48
What is a <u>periodic market</u>?	Local village markets that operate at regular intervals every few days; they attract nearby buyers and sellers, some of whom follow the market's shifting circuit.	448-49

EAST AFRICA	EAST AFRICA	
Which countries comprise this region?	Tanzania, Kenya, Uganda, Rwanda, Burundi, and southwestern Ethiopia.	449
Describe East Africa's natural environment.	Plateau, savanna country that becomes a steppe in the drier northeast; volcanoes and rift valleys mark the surface; Lake Victoria is a key feature.	449
Why is Tanzania viewed as an important lesson for Africa?	Despite limited resources and internal cultural divisions, it has achieved political stability through self-help African socialism.	450-51

How does Kenya's development differ spatially from Tanzania's?	Kenya's development is concentrated in the core area around Nairobi; Tanzania's is much more dispersed.	452-53
What is Uganda's current status?	The struggle to emerge from the chaotic Amin years continues as a shaky economy inhibits recovery and political instabilities are recurrent.	454
EQUATORIAL AFRICA	EQUATORIAL AFRICA	
Which countries make up this region?	Cameroon, Central African Republic, Zaire, Congo, Cabinda, Gabon, Equatorial Guinea, and southern Sudan. Chad is often included too (see map on p. 445), although that country was treated with the African Transition Zone in Chapter 6.	454-55
What is Zaire's leading physical feature?	The vast basin of the Congo (now Zaire) River, whose mineral wealth is concentrated within the basin's southeastern rim.	455
Why has the economy of Cameroon been more successful than those of its neighbors?	The Germans and then the French treated the country more constructively, leaving a good commercial agricultural base.	456
SOUTHERN AFRICA	SOUTHERN AFRICA	
Which countries are included in this region?	South Africa, Lesotho, Swaziland, Mocambique, Zimbabwe, Zambia, Malawi, Angola, Namibia, and Botswana (Madagascar is technically included as well).	456
What is the significance of the Copperbelt?	This zone of copper and other rich mineral deposits is Zambia's portion of Africa's greatest resource region--which stretches south from southeastern Zaire through Zambia, Zimbabwe's Great Dyke, and into the heart of northern South Africa.	456-58
Who are the two main peoples of Zimbabwe?	The Shona-speaking Mashona, and Nguni-speaking Ndebele.	460

MAP EXERCISES

MAP COMPARISON

1. Many of Africa's countries contain diverse cultural groups that were largely thrown together by the Europeans in the political territories that were carved out in the colonial era. Compare the maps on pp. 419, 439, and 441 and record your observations about this fragmented pattern, its evolution over the past century, the countries which are most strongly affected, and how recent population movements in the independence era are adjusting some established cultural-geographic patterns.

2. Compare and contrast the maps on p. 436 with the map on p. 51. Why did the trans-Atlantic slave trade become such a lucrative operation, achieving mass proportions in such a short time period? Why did West Africa spawn so prominent a slave trade? How does the present social geography of the Caribbean and northeastern South America reflect the aftermath of the Atlantic slave trade (refer back to appropriate sections of Chapters 4 and 5)?

3. The diagram of colonial transport development (p. 442) describes 6 stages of network growth through the end of the colonial era. Compare the generalized pattern in the final stage (F) with the current road and railroad patterns in the four regional maps (pp. 446, 451, 455, and 457), and describe the changes that have occurred since independence (if any).

4. Carefully study the map of South Africa on p. 466 and search for patterns that reveal segregation among ethnic/racial groups. Make observations on the overall patterns of apartheid, noting those cities and subregions indentified with various ethnic groups.

MAP CONSTRUCTION

(Use outline maps at the end of this chapter)

1. In order to familiarize yourself with African physical geography, place the following on the first outline map:

    a. Rivers: Niger, Senegal, Volta, Benue, Zaire (Congo), Lualaba, Ubangi, Kasai, Blue Nile, White Nile, Zambezi, Cubango, Limpopo, Vaal, Orange

    b. Water bodies: Gulf of Guinea, Mocambique Channel, Lake Chad, Lake Volta, Lake Victoria, Lake Turkana, Lake Tanganyika, Lake Albert, Lake Malawi, Lake Kariba

    c. Land areas: Madagascar, Futa Jallon Highlands, Ethiopian Highlands, Drakensberg, Sahara Desert, Sahel, Kalahari Desert, Namib Desert

2. On the second map, political information should be entered as follows:

   a. Label each country with its name

   b. Locate and label each capital city with the symbol *

3. On the third outline map, urban-economic information should be entered as follows:

   a. Cities (locate and label with the symbol .): Dakar, Bamako, Niamey, Conakry, Freetown, Monrovia, Ouagadougou, Abidjan, Kumasi, Accra, Lome, Porto Novo, Kano, Ibadan, Lagos, Enugu, Port Harcourt, Maiduguri, Ndjamena, Nairobi, Mombasa, Kisumu, Kampala, Jinja, Dar-es-Salaam, Mwanza, Yaounde, Douala, Matadi, Kinshasa, Bangui, Brazzaville, Kananga, Lubumbashi, Luanda, Lobito, Lusaka, Ndola, Blantyre, Harare, Bulawayo, Beira, Maputo, Tananarive (Antananarivo), Johannesburg, Pretoria, Cape Town, Bloemfontein, Durban, Port Elizabeth, East London

   b. Economic regions (identify with circled letter):

      A - Copperbelt
      B - Great Dyke
      C - Witwatersrand
      D - Niger Delta
      E - Great Lakes Region
      F - Tan-Zam Corridor
      G - Transvaal
      H - Highveld

PRACTICE EXAMINATION

SHORT-ANSWER QUESTIONS

Multiple Choice

1. The edge of the African plateau, especially in Southern Africa, is known as the:

   a) Rift Valley
   b) Gondwana Mountains
   c) Great Escarpment
   d) Great Dyke
   e) Great Rift

2. The country in the process of moving its capital from Lagos to the new centrally-located city of Abuja is:

   a) Nigeria
   b) Tanzania
   c) Zaire
   d) Burkina Faso
   e) Ethiopia

3. The geographic concept associated with the successive cultural imprints left on a region's landscape over time is:

   a) underdevelopment
   b) contagious diffusion
   c) sequent occupance
   d) acculturation
   e) culture realm

True-False

1. Before independence, the modern state of Zaire was a colony of France.

2. The upper basin of the Niger River has been an area of far greater cultural development over time than its delta region.

3. The Copperbelt is partially located in the country formerly known as the Belgian Congo.

Fill Ins

1. The Ibos and Yorubas inhabit the country called ______________.

2. Zimbabwe's greatest concentration of mineral deposits are located in a corridor, which passes through the country's core area, called ______________.

3. The worst and most widespread of Africa's endemic diseases is ______________.

Matching Question on African Countries

_____	1.	Hausa-Fulani	A.	Tanzania
_____	2.	Former Portuguese colony	B.	Zaire
_____	3.	Former French Colony	C.	Equatorial Guinea
_____	4.	Kikuyu culture	D.	Kenya
_____	5.	Boers	E.	Zimbabwe
_____	6.	Formerly Northern Rhodesia	F.	Uganda
_____	7.	Dar-es-Salaam	G.	Mocambique
_____	8.	Mashona culture	H.	Nigeria
_____	9.	Former Spanish colony	I.	Senegal
_____	10.	Idi Amin	J.	South Africa
_____	11.	Desert country	K.	Zambia
_____	12.	Belgian Congo	L.	Botswana

ESSAY QUESTIONS

1. Define and discuss the concept of sequent occupance as it relates to historical cultural geography, and apply the idea to the region of the Upper Niger basin during the times of the empire of ancient Ghana, the period following the arrival of Islam, and the 1980s.

2. Discuss at some length the concerns of contemporary medical geography, and apply its methodology to the spatial pattern of a major disease that affects Africa today.

3. The continental drift hypothesis has been demonstrated to be a fairly powerful explanation of African hydrography (surface water patterns) and physiography. Show how one can use this approach to understand Africa's dominantly plateau-like terrain, the absence of mountain chains, the rift valleys, and the unusual courses of the realm's major rivers.

4. Perhaps no region has been more thoroughly disrupted and transformed by the colonial experience than West Africa. Discuss the major European penetrations here, the kinds of political territories that emerged, and how the region's disparate populations came into conflict during Nigeria's brutal civil war in the late 1960s and early 1970s.

5. The possibility of a "Green Revolution" for Africa is a topic of wide discussion today. Discuss the present condition of Subsaharan Africa's agriculture, the nature of the Green Revolution itself, and chances of success if it were widely implemented in this realm.

TERM PAPER POINTERS

The "Term Paper Pointers" section of Chapter 1 in this Study Guide offered suggestions about approaching research and writing on geographic realms and their components, and you may wish to consult this material if you are undertaking a report on an African region.

The enormous variations that occur across African space make a regional geography almost indispensable: besides the standard surveys (Best & de Blij, Grove [1978], Hance, and Senior & Okunrotifa) there are also systematic overviews that are quite useful (Bell, Boateng, Christopher [1984], Knight & Newman, and Mountjoy & Hilling). Individual regions are also covered in Crush, Gugler & Flanagan, Morgan [1976], Onyemelukwe & Filani, Smith, and Udo [both sources], and Harrison-Church--and Gourou is always a source to be consulted on tropical regions. Individual countries are surveyed by Frankel, Gade, N. Miller, Morgan [1985], and Yeager. Good atlases are Griffiths, Murray, and Ajayi & Crowder. The Time article [1984] offers updates on African trends since 1980, as do Lamb, Oliver & Crowder, Ungar, and Willis. The history of Africa is treated by Murdock, Fage & Oliver, Christopher [1984], Ajayi, Betts, Curtin, Fetter, Trimmingham, and Hibbert. Development and modernization in contemporary Africa is dealt with by Gould, O'Connor [1978], Berg & Whitaker, Bernard, Clarke et al., Hoyle, and Knight. Economic geography is covered in Martin & de Blij, Smith, Taaffe et al., Barwell et al., and Beyer. Population issues are reviewed by Clarke & Kosinski and Goliber. Urban geography is highlighted by O'Connor [1983], Obudho, and de Blij. Political geography is treated by Graubard, Nwafor, Boateng, and Christopher [1985]. The postindependence era is covered in "Africa's Woes," Boateng, Graubard, Lamb, Meredith, and Oliver & Crowder. The food crisis and related problems are treated in Berardi, Dumont & Mottin, Franke & Chasin, both Harrison sources, and Scott. Physical topics are discussed in Grove [1985] and Pritchard.

The medical geography Systematic Essay can also be used as a springboard for a research paper. Good introductions to the subject are found in Meade et al., Jones & Moon, Pyle, Howe, Learmonth, May, Gesler, and all McGlashan sources, and the atlas by Rodenwaldt & Jusatz is a landmark work. Applications to Africa are very well handled in Good, Hunter, Prothero, Stock, and Kloos & Thomson. Since many studies in medical geography focus on the spreading of diseases, it is well worth reviewing the material on spatial diffusion in Chapter 6 of the textbook; the Special Exercise keyed to that chapter in this Study Guide offers advice about finding and using spatial-diffusion data.

SPECIAL EXERCISE

The model of colonial transport development (Model Box 5, pp. 442-443) offers some valuable lessons about Third World modernization, and this exercise will lead you more deeply into the subject. To begin with, you should locate and read the article on which the Model Box is based: Edward Taaffe, Richard Morrill, & Peter Gould, "Transport Expansion in Underdeveloped Countries," Geographical Review, 53 (October 1963), 503-529.

Once you have read this article, find and read the following: Peter Rimmer, "The Changing Status of New Zealand Seaports, 1853-1960," Annals of the Association of American Geographers, 57 (March, 1967), 88-100. Rimmer extends the Taaffe-Morrill-Gould model, and elucidates on complications that can occur in the real world to impede the ideal-typical sequence postulated in the model. Compare the two studies, contrasting the West African and New Zealand situations. What lessons obtained from the New Zealand study might be reapplied to West Africa to better understand its transport-network evolution? If the West African research were being conducted anew today, what additional data would be needed to update the study to the present?

As an optional additional exercise, find and read another application of the ideal-typical network model--this one covering the growth of the U.S. transportation system since the early 19th century, undertaken by Edward Taaffe alone. His study appears in his book, Geography of Transportation, coauthored with Howard Gauthier and published by Prentice-Hall in 1973; the application of the model is found on pp. 47-52, with further elaboration on pp. 52-72. If you cannot locate this book, Taaffe has written a brief overview as a chapter in another work: Problems and Trends in American Geography, edited by Saul B. Cohen, and published by Basic Books in 1967 (see pp. 15-25).

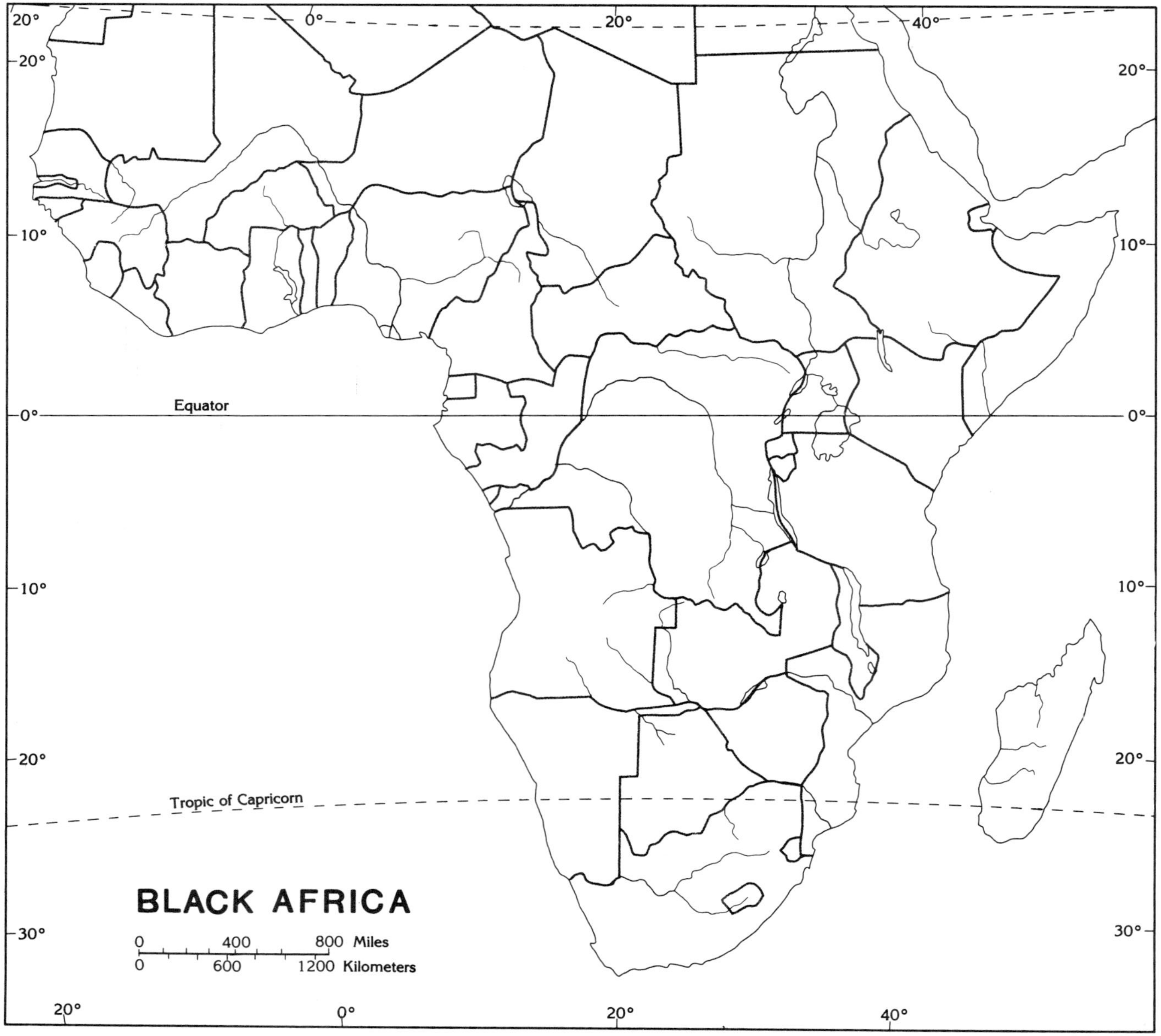

BLACK AFRICA
0 400 800 Miles
0 600 1200 Kilometers
Equator
Tropic of Capricorn
20°
0°
20°
40°
20°
10°
0°
10°
20°
30°

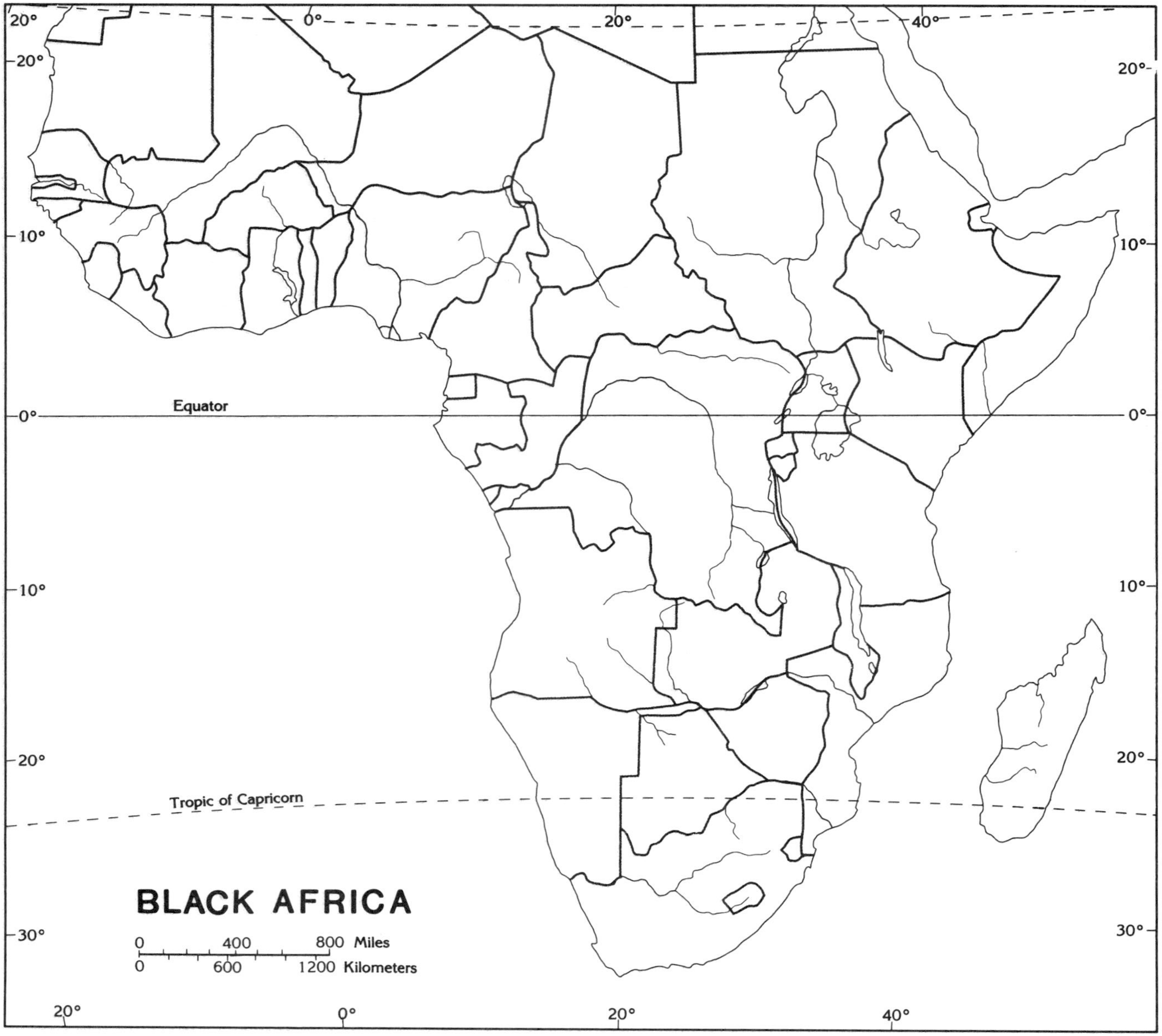
BLACK AFRICA
0 400 800 Miles
0 600 1200 Kilometers
Equator
Tropic of Capricorn
20°
0°
20°
40°
10°
0°
10°
20°
30°

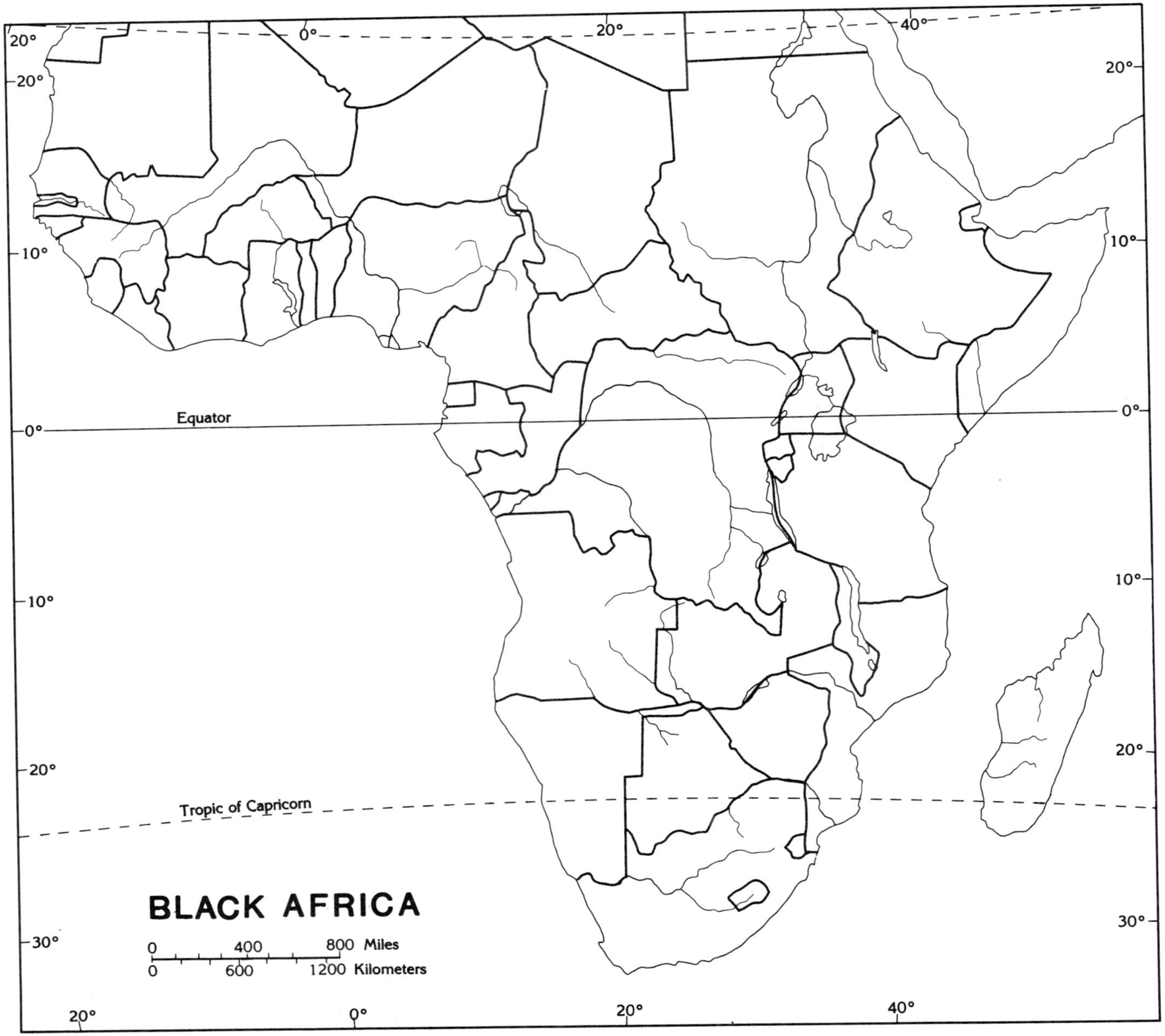
BLACK AFRICA
0 400 800 Miles
0 600 1200 Kilometers
Equator
Tropic of Capricorn
20°
0°
20°
40°
10°
0°
10°
20°
30°

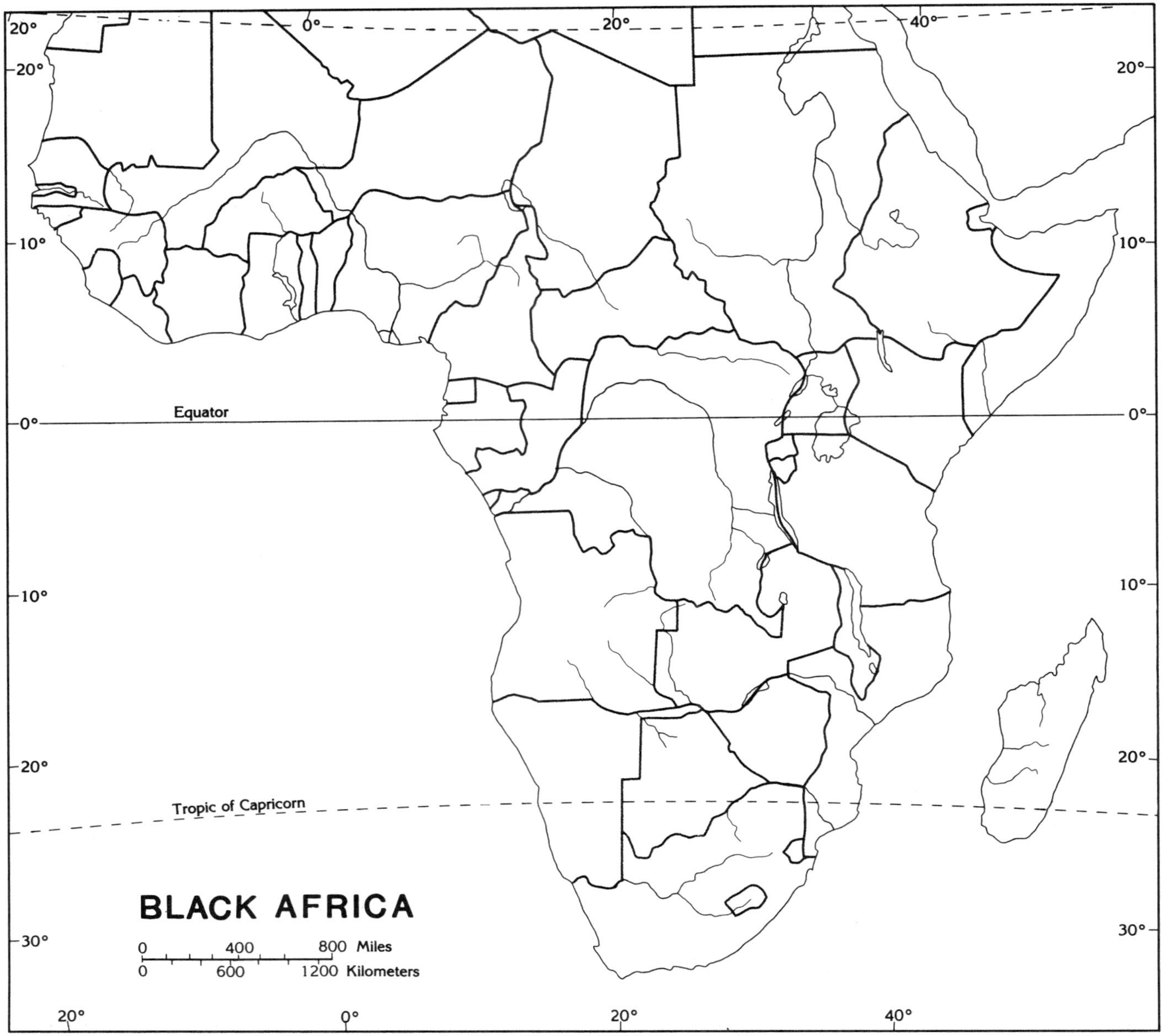
Equator
Tropic of Capricorn
BLACK AFRICA
0 400 800 Miles
0 600 1200 Kilometers
20°
0°
20°
40°
10°
30°

VIGNETTE: SOUTH AFRICA -- CHALLENGE AND CHANGE

OBJECTIVES OF THIS VIGNETTE

This Vignette focuses on the Republic of South Africa, which contrasts sharply with the other states of South Africa and the realm as a whole (see box on p. 465). This troubled country has experienced constantly increasing racial tensions in the 1980s, and its new treatment in this edition as a separate Vignette underscores the growing isolation of South Africa in the international community as well as the rising concern about its future. The Vignette opens with an introduction to South Africa's current problems, and outlines the historical developments that have produced one of the realm's most heterogenous--and polarized--populations. Next comes an overview of the country's physical geography, emphasing the considerable variations in the natural environments against which today's dramatic events are unfolding. An overview of economic geography follows, highlighting the mineral, agricultural, and infrastructural developments that have shaped South Africa's livelihood patterns and economic growth since the mid-nineteenth century. The final section deals with political geography, tracing the rise of the present white-minority governmental system, the racist policies of the past four decades and their impacts, and how the country's potentials are now being lost amid the rising turmoil of the post-1985 period.

Having learned the regional geography of South Africa, you should be able to:

1. Appreciate its place in Subsaharan Africa and on the current world scene.
2. Describe the broad dimensions of its varied human and physical geography.
3. Understand the importance of South Africa's rich mineral-resource base and the potentials vis-a-vis its further development.
4. Understand the key role the country has played in the development of Southern Africa since the beginning of the colonial era.
5. Differentiate among the major black, white, Asian, and racially-mixed groups that are the major players on South Africa's stage today, and the socio-geographical patterns they exhibit.
6. Trace the historical geography of the country's development, particularly the progress of the Dutch-heritage Boers and the rise of Afrikaner political power.
7. Understand the present politico-geographical situation in South Africa, especially the racial polarization that has intensified since the Afrikaners came to rule in 1948.
8. Locate the leading physical, cultural, and economic-spatial features of this country on an outline map (a skill already tested in the preceding chapter of this Study Guide).

GLOSSARY

Buffer Zone (464)

A group of countries separating ideological or political adversaries. In the case of South Africa and Namibia, the perimeter of states just to the north (Angola, Zambia, Zimbabwe, and Mocambique) that separated the southern end of the continent from the full thrust of African nationalism. When the colonial era finally ended in this zone during the 1970s, South Africa became a "front-line" state vis-a-vis the advancing tide of independence.

Afrikaners (465)

The descendants of the original Dutch colonists of South Africa, known also as the Boers, who steadily gained numerical and political strength until they came to power in 1948.

Coloured People (465)

The mulatto population resulting from the intermarriage of European settlers and blacks in the Cape area. Today, this mixed-ancestry group forms the majority population in and around Cape Town.

Veld (467)

Open grassland on the South African plateau, becoming mixed with scrub vegetation at lower elevations where it is called bushveld. As in Middle and South America, an altitudinal zonation exists in the form of highveld, middleveld, and lowveld.

Apartheid (470)

South Africa's policies of racial separation that underlie a totally segregated society.

Separate Development (470)

The new name for South Africa's apartheid scheme, which sought to allocate all racial groups into their own "homelands." Although supposedly "parallel" in its aims, nonwhites were assigned the country's poorest resources and least productive areas.

Bantustan (470)

The popular name of the separate development program, which was designed to redistribute all racial groups to their own "homelands."

Polarization (471-472)

A population divided into hardened factions as an internal crisis intensifies. In South Africa today, the leading factions are the military, the conservative white-minority government, the institutional opposition, disenfranchised blacks, and supporters of exiled or jailed rebels.

SELF-TESTING QUESTIONS

Cover this side of the page with a sheet of paper. Uncover each line after you have attempted to answer the question in the left column. If necessary, refer to textbook page(s) listed at the right.

Question	Answer	Page
SOUTH AFRICA'S CHARACTERISTICS	SOUTH AFRICA'S CHARACTERISTICS	
What are the country's general geographical dimensions?	It is large and varied in its natural environments, and contains about 35 million people--a population 2-1/2 times as large as Southern Africa's next largest country.	464
Why is South Africa now a "front-line" state vis-a-vis the wave of African independence?	The surrounding buffer zone of states (except for the South African colony of Namibia) have now all gained independence.	464
Who were the Boers?	The descendants of the original Dutch settlers, who were driven off from the Cape area by British colonists, but who eventually emerged in the north as the <u>Afrikaners</u> to take control of the government in 1948.	465
Name the leading ethnic groups in South Africa today.	Besides the Afrikaners and British-heritage whites, there is a large number of black peoples (including the Zulu, Xhosa, and Sotho), the Cape Coloured, and the Indians (South Asians) of Natal Province.	465-66
What is the <u>veld</u>?	The open grassland of the South African plateau, stronghold of the Afrikaner population. Vertical zonation is discernible in the <u>highveld</u>, <u>middleveld</u>, and <u>lowveld</u>.	467
ECONOMIC GEOGRAPHY	ECONOMIC GEOGRAPHY	
How did the discovery of diamonds open up the South African interior?	The finds around Kimberley in the 1860s attracted thousands of fortune hunters and laborers as well as railroad construction and thriving mining/manufacturing industry.	468
How did the gold discoveries affect Boer political life?	The major goldfields were located in the core of Boer territory and also lured thousands of outsiders. The British fought a short war and defeated the Boers just after 1900. Eventually, their Afrikaner successors regrouped and became political leaders after 1948.	468

Question	Answer	Text page
Besides diamonds and gold, name the other assets that supported South Africa's economic development over the past century.	The favorable natural environment yielded a variety of fruits, grains, and vegetables; cheap labor was plentiful; several other rich mineral deposits were exploited, including iron, copper, nickel, tin, and chromium; a major metal fabrication industry arose; a superb transport network matured; foreign investment flowed in and white immigration continued (at least through the 1970s); thriving cities mushroomed.	468-69
What are the country's leading commercial agricultural activities?	Wheat (especially in the highveld Maize Triangle), tropical fruits, sugar, grapes, apples, and sheepraising.	469
POLITICAL GEOGRAPHY	POLITICAL GEOGRAPHY	
Was the country developed equitably or did regional economic differences arise?	Internal differences were always notable and have intensified in recent decades. The great cities, booming industries, and modern farms resemble a developed country; the remainder of South Africa, overwhelmingly nonwhite, exhibits the miserable conditions typical of Third World Subsaharan Africa (see photo on text p. 470).	469
Why was 1948 so important a political turning point in South Africa?	It was then that the Afrikaners finally came to outnumber other white peoples; they elected a conservative government that quickly institutionalized racism under the policies of <u>apartheid</u>.	469-70
What is <u>apartheid</u>? What is its successor, <u>separate development</u>?	South Africa's unforgiving racial subjugation and segregation policy that officially separated nonwhites from whites; in recent years, the policy came to be called <u>separate development</u> and sought to isolate each racial group in its own "homeland" (in the process giving majority blacks the worst land and poorest resources).	470
Why has the world's attention been turned to South Africa in the 1980s?	The black opposition (spurred on by the African National Congress) began a major struggle; the "homeland" creation program began to flounder; economic growth slowed markedly as commodity prices changed on the world market; a growing number of Western countries "disinvested" and ceased to trade with South Africa to protest the white-minority government's increasingly offensive racist policies.	471

What are the prospects for South Africa in the immediate future?	Discouraging, because the country's potentials are being lost as polarization heightens and a climactic confrontation draws closer as the government is forced to use all its capacities simply to keep control.	472

MAP EXERCISE

There is no map exercise in this Vignette: South Africa was included in the map activities of the previous chapter.

PRACTICE EXAMINATION

SHORT-ANSWER QUESTIONS

Multiple Choice

1. South Africa's South Asian population is known as the:

   a) Cape Coloured
   b) Durbans
   c) Afrikaners
   d) Zulu
   e) Indians

2. The stronghold of the Afrikaner population is located in the physiographic zone called the:

   a) highveld
   b) lowveld
   c) Drakensberg
   d) Cape Ranges
   e) White Belt

True-False

1. South Africa is Africa's sole remaining state in which minority rule is exercised by people of European descent.

2. The Afrikaners conquered the Boers during World War II, thereby assuring that the long-time apartheid policy would be defeated at the next national election.

Fill-Ins

1. Today's Afrikaners are the successors to a colonist population that originated in the European country of ______________.

2. "Bantustans" or "homelands" are cornerstones of the white-minority government policy known as ______________, which has succeeded apartheid in recent years.

Matching Question on South African Cities

_____	1.	Administrative capital	A.	Cape Town
_____	2.	Infamous capital of a "homeland"	B.	Pretoria
_____	3.	Orange-Vaal confluence focus	C.	Durban
_____	4.	Stronghold of South Asian population	D.	Soweto
_____	5.	First white settlement	E.	Sun City
_____	6.	One of the largest black townships	F.	Johannesburg
_____	7.	Largest city in South Africa	G.	Kimberley

ESSAY QUESTIONS

1. Racial-separation policies in South Africa have been shown to divide its population into a far more complex residential pattern than mere white/black segregation. Review the country's current racial policies, the geographical shaping of a multi-racial society, and the trend toward "homelands" for each major ethnic group.

2. The Afrikaners have exerted political control over South Africa for more than four decades and are grimly determined to remain in power as the confrontation with the nonwhite majority population intensifies. Trace the origins of the Afrikaners, their early movements within the country, the establishment of their territorial stronghold, their rise to national political power, and their responses to the gathering storm of protest in the 1980s against Afrikaner leadership from both inside South Africa and from abroad.

TERM PAPER POINTERS

The literature cited on p. 473 offers a variety of materials on South Africa, many of them quite current; you are further reminded that much is also available in the general sources on Subsaharan Africa provided on pp. 461-463. The volatility of the South African situation assures its continuing prominence in the news during the late 1980s and the opening of the 1990s. A recent overview is the cover story that appeared in Time in the May 4, 1987 issue ("South Africa: Hints of Hope--Afrikaners Begin to Unbend," pp. 28-38). Lengthier accounts of the contemporary scene can be found in the sources by Blumenfeld, Leach, Lelyveld, and Smith [1987] on text p. 473. Your instructor and librarian should also be consulted in the search for the most up-to-date materials at the time of your research.

There is no Special Exercise for this Vignette

CHAPTER 8: SOUTH ASIA -- RESURGENT REGIONALISM

OBJECTIVES OF THIS CHAPTER

South Asia is the world's second largest population concentration, with slightly over 1 billion people residing in 1988 in India, Bangladesh, Pakistan, Sri Lanka, Nepal, Bhutan, and the remaining microstates. The Systematic Essay--geomorphology--is particularly well suited to this realm, which contains some of the most dramatic topography on earth including the highest mountain range. Following the introduction, the rich historical geography of South Asia is traced, highlighting its significant role as a hearth of civilization and its transforming colonial experience under the British. The political and economic geography of India during the post-1947 independence period is reviewed, emphasizing progress toward the modernization of its highly traditional society. The country's immense demographic problems are highlighted next, and are used as a basis to explore the world population crisis (which is no longer at its worst in this realm). The chapter closes with surveys of Bangladesh, Pakistan, Sri Lanka, Nepal, and Bhutan, treating geographic dimensions as well as current problems faced by each country.

Having learned the regional geography of South Asia, you should be able to:

1. Understand the basic concepts of geomorphology and their significance for interpreting the natural environment.
2. Describe the physiography and the wet-monsoon dynamics of the Indian subcontinent.
3. Trace the major stages of the realm's historical geography, particularly the dramatic transition from colonialism to independence.
4. Understand the unique federal political structure of India and its underlying cultural mosaic.
5. Trace the recent geography of India's economic development as it has reshaped agriculture, manufacturing, and urbanization.
6. Appreciate the crisis of the realm's population numbers, and its relationship to demographic cycles in the Third World.
7. Understand the particular miseries of life throughout Bangladesh, and the bleak prospects that this country faces in the foreseeable future.
8. Understand the somewhat greater opportunities for progress that exist in Pakistan, but are tempered by internal and external political problems that could explode at any time.

9. Understand the forces that have helped Sri Lanka to make some recent gains, which are now threatened by the dangerous conflict between the majority Sinhalese and the minority Tamils.

10. Locate the major physical, cultural, and economic-spatial features of the realm on an outline map.

GLOSSARY

Geomorphology (478)

The geographic study of the configuration of the earth's surface.

Tectonic forces (478)

Produce internal crustal movements that shape surface landforms; these can be vertical or horizontal movements, and are understood today in terms of the plate tectonics theory discussed on text pp. 9, 425-426.

Erosion (478)

The removal of weathered rock and its further breakdown by such agents as running water, glaciers, wind, and ocean waves.

Igneous rocks (479)

Rocks formed by the cooling of volcanic materials.

Sedimentary rocks (479)

Rocks formed by the deposition and subsequent solidification of loose eroded material.

Metamorphic rocks (479)

Rocks formed from the altering of preexisting igneous or sedimentary rock by the reintroduction of substantial heat and pressure.

Folding (479)

Warping or bending of rock layers into "folds," produced by the heat and pressure of crustal deformations that make solid materials plastic.

Faulting (479)

Fracturing of rocks produced by extreme crustal deformations, often triggering earthquakes; can occur vertically (as in Africa's rift valleys) or horizontally (as in California's San Andreas Fault).

Caste system (481)

The strict social-class segregation of people, in Hindu society, on the basis of ancestry and occupational status.

Wet monsoon (482-483 box)

The rainy season produced by an onshore airflow that dominates for months, occurring in the hot summer months when low pressure over the land sucks in moisture-laden air from the adjacent cooler ocean; especially pronounced throughout coastal Asia.

Partition (486-487)

The division of British India, upon independence in 1947, into India and the two Pakistans along religious lines, with Hindus dominant in the former and Moslems in the latter two.

Superimposed political boundary (487-488)

A boundary placed on a landscape without regard for existing cultural patterns.

Centrifugal forces (491)

Disunifying or divisive political forces within a state; examples are internal religious conflicts, racial strife, and contrasting regional outlooks.

Centripetal forces (491 box, 492)

Unifying forces that bind a state together; examples are charismatic leaders, perceived external threats, and widespread commitment to an existing governmental system.

Physiologic density (492)

The number of people per unit area of cultivable or arable land.

Green Revolution (493, 431 box)

The introduction of higher-yield strains of rice and wheat throughout much of non-communist Asia in the 1960s, that improved agricultural productivity; unfortunately, overall gains in living standards were minimal.

Doubling time (498-499)

The length of time required for a population to double its size; at present growth rates, the world's population will double only 41 years from now.

Malthusian (501 box)

Adherent of the point of view taken by the British economist Thomas Malthus in 1798 that food supply increases could not match population increases, thereby resulting in famines and the ultimate cessation of population growth.

Typhoon (504)

A tropical cyclonic storm or hurricane with sustained wind speeds in excess of 74 m.p.h.

Forward capital (509, 351)

Capital city located in contested territory near a border, confirming a state's desire to maintain its presence in that area.

SELF-TESTING QUESTIONS

Cover this side of the page with a sheet of paper. Uncover each line after you have attempted to answer the question in the left column. If necessary, refer to textbook page(s) listed at the right.

Question	Answer	Page
GEOMORPHOLOGY	GEOMORPHOLOGY	
What is geomorphology?	The geographic study of the configuration of the earth's surface.	478
What are tectonic geomorphic processes?	Those triggered by internal crustal movements, with each process shaping its own particular assemblage of landforms.	478
What is meant by the terms <u>weathering</u>, <u>erosion</u>, and <u>deposition</u>? What is the difference between degradation and aggradation?	Rock disintegration by temperature change and chemical action; the removal of weathered rock and its further breakdown by such erosional agents as water, ice, and wind; accumulation of transported eroded materials, such as in the lower portion of a river valley. <u>Degradation</u> refers to weathering and erosion; <u>aggradation</u> refers to depositional accumulations.	478
What are the three major types of rocks and their origins?	<u>Igneous</u> rocks are associated with volcanism; <u>sedimentary</u> rocks originate from deposition and subsequent lithification; <u>metamorphic</u> rocks are formed from altered igneous or sedimentary rocks.	479
What is the difference between folding and faulting?	Folded rock layers are produced by heat that makes solid materials plastic and "bends" or "warps" the rock beds; faults are rock fractures caused by even greater tectonic pressures, often creating earthquakes.	479
SOUTH ASIAN CHARACTERISTICS	SOUTH ASIAN CHARACTERISTICS	
Which countries comprise this realm?	India, Pakistan, Bangladesh, Sri Lanka, Nepal, and Bhutan.	475

What became of the two Pakistans, created by the British before granting independence to the Indian subcontinent in 1947?	These two Moslem states, deliberately created to be separate from Hindu-dominated India, went their separate ways after their 1971 war; West Pakistan thus became Pakistan, and East Pakistan became Bangladesh.	475

PHYSIOGRAPHY

Describe the realm's three major physical regions.	The northern mountain belt, the southern (Deccan) plateau, and the intervening riverine lowlands.	476
What is the North Indian Plan?	The belt of alluvial lowlands stretching between the Indus on the west to the Brahmaputra on the east; the Ganges is its major river.	477
What is the Punjab?	The low hills between the North Indian and Indus Plains, the so-called "land of five rivers" astride the India-Pakistan border.	477
What are the boundaries of the Deccan?	On the north, the Vindhyas and the Tapti & Godavari Rivers; on the west, the Western Ghats lining the Malabar-Konkan Coast; on the east, the Eastern Ghats paralleling the Coromandel-Golconda Coast.	480

HISTORICAL GEOGRAPHY

Who were the Aryans? What was their impact on the emergence of India's modern culture?	Invaders from western Asia who conquered the early Indus Valley civilization after 2000 B.C., but who adopted many of its innovations and pushed settlement frontiers east into the Gangetic Plain and south into the center of the peninsula. India's culture developed from this beginning, including the Hindu religion and its rigid caste system.	481
What lasting impact did Buddhism make?	It was dominant during the Mauryan Empire (3rd century B.C.-2nd century A.D.), but soon declined in South Asia only remaining strong in Ceylon (Sri Lanka) where it still prevails; Buddhism today is mainly centered in East and Southeast Asia (see Fig. 6-3).	481-82
What was the lasting impact of Islam?	After the 10th century it was a strong influence, driving out Buddhism but not Hinduism, which remained dominant in the Ganges core area; Moslems remain a sizable minority (11%) in India--and form the major cultural bases in Pakistan and Bangladesh.	483

Describe the realm's colonial-era experience.	The British quickly emerged as the dominant colonial power, first through the East India Company, and then outright political control from 1857 to 1947. The British introduced many innovations, but forced the colonial economy to become a raw-material producer subservient to the English master.	484-86
Describe the events surrounding the 1947 partition of British India.	The independence movement led by Gandhi is a well-known story, and the British left in 1947; however, before withdrawing they separated their former territory into Hindu-dominated India and the two Islamic Pakistans, thereby inducing mass migrations, conflict, and social stresses that still exist.	486-88

POLITICAL GEOGRAPHY

POLITICAL GEOGRAPHY	POLITICAL GEOGRAPHY	
Describe India's federal political structure.	This most populous federal state has 23 internal states and 7 territories with boundaries largely drawn to respect linguistic differences.	488
What is India's present linguistic status?	Hindi is the largest of the 14 "official" languages, spoken by about 35% of the people (there are hundreds of local languages); English has emerged as the *lingua franca*, the chief language of the business world.	488-89
What are *centrifugal* and *centripetal* forces? How do they operate in India today?	Disunifying and unifying forces, respectively. Current centrifugal forces are religious conflict, linguistic differences, and contrasting regional outlooks; centripetal forces include Hinduism and the support for national economic development.	491 box

ECONOMIC DEVELOPMENT

ECONOMIC DEVELOPMENT	ECONOMIC DEVELOPMENT	
Why is Indian agriculture so stagnant?	Because inefficient traditional farming methods are widespread, and because land continues to be divided into very small plots.	492-93
Has the Green Revolution improved farm productivity?	Yes, but the population explosion since the 1960s has prevented the closing of the gap between population and food supply.	493
How has industrialization progressed?	Slowly, despite good coal and iron deposits; investments continue but underdevelopment patterns are hard to break.	495-98

POPULATION GEOGRAPHY

Question	Answer	Page
What is the current doubling time for the world population?	About 41 years; the 4 billion humans alive in 1975 will double in size to 8 billion shortly after the year 2010.	498-99
What is meant by the term (crude) birth rate?	The annual number of live births per 1000 population.	500
Which regions of India are growing most rapidly? Most slowly?	As the map on p. 500 shows, the states adjacent to Bangladesh exhibit the highest growth rates; the south, Punjab, and lower Gangetic Plain, the least.	500-1
Is the Demographic Transition Model applicable to India?	Probably not: its population is already much too large, the raw-material base too small, and the opportunity for mass emigration nonexistent.	502
Has the rate of urbanization been as rapid in India as elsewhere in the Third World?	Surprisingly not, although the pace is quickening in the 1980s; yet, over 200 million (25% of the population) do reside in urban areas, most in overcrowded communities that can hardly be expected to absorb the additional millions now arriving annually in the cities--a crisis that has just begun to be dealt with. Enormous social contrasts necessitated by the caste system (photo p. 503) further exacerbate the coming urban crisis.	502

BANGLADESH

Question	Answer	Page
Describe the country's agricultural situation.	Fertile alluvial soils permit highly intensive farming, tea and jute for cash, rice for subsistence.	503-4
Why is this country the epitome of Malthusian forecasts?	It is in a constant state of hunger and local starvation, dislocation, and daily misery.	504

PAKISTAN

Question	Answer	Page
How does this country's environment differ from the rest of South Asia's?	This is a desert and steppe country, and water shortages are a chronic problem in the settled zones of Pakistan.	506-7
Where is the Moslem stronghold here?	In the Pakistani portion of the Punjab, especially around Lahore near the Indian border.	507-8

What are some examples of irredentism here?	Baluchistan in the arid Southwest, and the Pakhtuns along the Afghani border in the northwest; for good measure, the Kashmir dispute further compounds the country's centrifugal forces.	508-9
SRI LANKA	SRI LANKA	
Which two groups are in conflict here?	The Buddhist, Aryan, Sinhalese majority and the Dravidian, Hindu, Tamil minority.	510

MAP EXERCISES

MAP COMPARISON

1. Compare the world distributions of population growth (Fig. 8-18, pp. 498-499) and life expectancy at birth (Fig. 7-10, pp. 428-429). What relationship, if any, exists between national population growth and life expectancy (offer global observations as well as brief observations for each realm [map of world realms is found on pp. 42-43])? Can your conclusions be extended to include population densities (select 30 countries at random from Appendix A on pp. 611-614 to arrive at a determination)?

2. Compare the physiographic maps of South Asia and Africa (pp. 476 and 423), noting key similarities and differences. Describe India's position within the continental drift scheme (Fig. 7-8, p. 425) and its present situation within the tectonic-plate framework (Fig. I-6, p. 12).

3. The linguistic map of India is a result of its cultural evolution over the past 3000 years. Analyze that map (p. 486), accounting for its major differences within the broad framework of the historical-geographic forces that shaped the subcontinent's division into Indo-Aryan and Dravidian cultural spheres.

4. Environmental conditions are frequently associated with crop concentrations in subsistence economies. Compare the map of India's agriculture (p. 494) with the distributions of its climates (p. 19), vegetation (p. 23), and soils (p. 25), and make observations about the relative productivity of India's regions and the associations of the following crops with specific climate types: cotton, rice, and wheat. Do your conclusions also hold for Sri Lanka (map p. 510)?

5. To get a better idea of the growth rate of each of India's regions, list 5 countries in other realms that belong to each of the color categories shown in the legend of the map on p. 500 (the world growth-rate map is shown on pp. 498-499).

MAP CONSTRUCTION

(Use outline maps at the end of this chapter)

1. In order to familiarize yourself with South Asian physical geography, place the following on the first outline map:

   a. Rivers: Indus, Ganges, Brahmaputra, Hooghly, Meghna, Godavari, Sutlej, Narmada, Cauvery, Mahanadi, Ghaghra, Yamuna, Tapti

   b. Water bodies: Bay of Bengal, Arabian Sea, Palk Strait, Ganges Delta, Rann of Kutch

   c. Land bodies: Sri Lanka, Sind, Kathiawar Peninsula, Cardamom Upland, Thar Desert, Deccan Plateau, Malabar Coast, Coromandel Coast, Konkan Coast, Golconda Coast, North Indian Plain, Punjab, Indus Plain, Baluchistan, Assam Uplands, Chota Nagpur Plateau

   d. Mountains: Himalayas, Hindu Kush, Karakoram Range, Western Ghats, Eastern Ghats, Vindhya Range, Khasi Hills, Sulaiman Range, Kirthar Range

2. On the second map, political-cultural information should be entered as follows:

   a. Draw in the internal state boundaries of India (text p. 489)

   b. Draw in the language regions mapped on p. 486 and color them in appropriate shades

   c. Compare the two and offer conclusions. As an optional additional exercise, consult an appropriate atlas, and map in concentrations of Moslem minorities within India; how do these clusters relate to linguistic patterns and how have they changed since 1951 (Fig. 8-9, right-hand map)?

3. On the third outline map, urban-economic information should be entered as follows:

   a. Cities (locate and label with the symbol •): Bombay, Calcutta, Madras, New Delhi-Delhi, Bangalore, Ahmadabad, Varanasi, Hyderabad (once each for India and Pakistan), Amritsar, Kanpur, Pune, Bhilai, Bhopal, Patna, Jammu, Srinagar, Agra, Chandigarh, Nagpur, Mysore, Pondicherry, Madurai, Jamshedpur, Lucknow, Jaipur, Karachi, Lahore, Rawalpindi, Islamabad, Multan, Peshawar, Dhaka, Chittagong, Khulna, Kathmandu, Thimphu, Gauhati, Colombo, Jaffna, Anuradhapura

   b. Economic Regions (identify with circled letter):

      A - Punjab
      B - Assam
      C - Bihar-Bengal Industrial Region
      D - Chota Nagpur Industrial Region
      E - Maharashtra-Gujarat Industrial Region
      F - Delhi Conurbation
      G - Nagaland
      H - Sind
      I - Baluchistan
      J - Hindustan
      K - Vale of Kashmir
      L - Pakhtunistan

PRACTICE EXAMINATION

SHORT-ANSWER QUESTIONS

Multiple Choice

1. The majority religion practiced in Sri Lanka is:

    a) Dravidian
    b) Aryan
    c) Islam
    d) Buddhism
    e) Hinduism

2. The partitioning of Hindu India from Moslem Pakistan occurred in:

    a) ca. 460 B.C.
    b) 1857
    c) 1947
    d) 1971
    e) 1984

3. Which of the following is not located in Pakistan:

    a) Deccan Plateau
    b) Sind
    c) Punjab
    d) Islamabad
    e) Baluchistan

True-False

1. The Pakhtun irredentist movement in northern Pakistan is based on cultural linkages to neighboring India.

2. Sikhism developed and is still based in the Punjab region.

3. Dravidians are in the majority in the population of the island-nation formerly called Ceylon.

Fill-Ins

1. The social stratification that dominates Hindu India is known as the ______________ system.

2. The lowland of far southeastern India that borders the Indian Ocean and contains the city of Madras, is called the ________________ Coast.

3. The amount of cultivated land per person is known as the ________________ density.

Matching Question on South Asia

_____	1.	Aryan Buddhist	A.	Sepoy Rebellion
_____	2.	Bihar-Bengal industrial region	B.	Malabar-Konkan Coast
_____	3.	Sikh religious capital	C.	Aurangzeb
_____	4.	Forward capital today	D.	East Pakistan
_____	5.	India's lingua franca	E.	Sind
_____	6.	Dravidian language	F.	Sinhalese
_____	7.	India takeover by British government	G.	Tamil
_____	8.	Holy city of Hindustan	H.	Islamabad
_____	9.	Indus Delta	I.	English
_____	10.	Mogul Empire	J.	Calcutta
_____	11.	Disastrous 1970 typhoon	K.	Varanasi
_____	12.	Western Ghats	L.	Amritsar

ESSAY QUESTIONS

1. The Kashmir problem continues to be an unsolved crisis of major political-geographic significance to South Asia. Trace the origin of this conflict, its evolution through the 1980s, and how the Soviet hegemony over nearby Afghanistan could affect this stalemate.

2. Bangladesh is undoubtedly plagued by more miseries than almost any other country in the world. Review the environmental, cultural, and economic forces that created this situation, and discuss the relationship between the country's population geography and hopes for future development and modernization.

3. Discuss the evolution of Hinduism as the leading religion and social system in India. Trace the various influences and challenges of the Mauryan Empire, the Mogul Empire, the British colonial era, and the recent course of Hinduism since independence.

4. Discuss the current agricultural geography of India. Make a sketch map that shows the regional distributions of rice and wheat, and discuss their relationship to the dynamics that shape the country's annual wet monsoon.

5. India's industrialization and urbanization have proceeded rather slowly over the past generation. Discuss the reasons for the leisurely manufacturing expansion, and why urbanization has accelerated markedly since 1980.

6. Pakistan is probably the most crucially-located state in all of Asia, standing between Soviet-controlled Afghanistan and both the Indian Ocean and teeming India. Provide an overview of Pakistan's internal geography, and speculate on its future as a strategic base for the non-communist West.

TERM PAPER POINTERS

The "Term Paper Pointers" section of Chapter 1 in this Study Guide offered suggestions about approaching research and writing on geographic realms and their components, and you may wish to consult this material if you are undertaking a report on a South Asian region.

Size and diversity of the realm as a whole, and within India in particular, makes a regional geography an indispensable reference. Farmer [1984] and Johnson [1983] are the newest sources, Spate & Learmonth a classic work, and Spencer & Thomas a broad overview; Dutt and Geib are the compilers of a comprehensive new atlas, and the Millers' bibliography is still useful. For India alone, the key works (in addition to those just mentioned) are Sopher [1980], Dutt [1972], Johnson [1979], Noble & Dutt, McLane, Cassen, and the magnificent historical atlas compiled by Schwartzberg (which also includes countries of the Indian perimeter). Pakistan is treated by Burki, Siddiqi, Feldman, and Singhal; Bangladesh by Johnson [1982], Baxter, Er-Rashid, and O'Donnell. As for the smaller countries, Sri Lanka is covered in DeSilva [both sources], Fernando, Hennayake & Duncan, and Tambiah; Karan focuses on Nepal [1960] and Bhutan [1967].

As for topical treatments, cultural geography is reviewed in Bhardwaj, Dumont, Hutton, Frazer, Stevens [May, 1983], Lukacs, Maloney, and Sopher [1967]. Political geography is covered in Allan, Beer & Jacob, Ganguly, Dikshit, and Stevens [August, 1983]; a splendid account of Pakistan's crucial strategic role may be found in the Wall Street Journal of October 2, 1984 under the title of "Pushing and Pulling Apart Pakistan" by Rosanne Klass. Historical geography is handled in Lall and Hutton, and various aspects of economic geography are covered in Barker, Bayliss-Smith, Farmer [1977], Hanson, Huke, Murton, Whyte, Berry, and Chakravati. Urban geography is treated in Dutt [1983], Brush, Kulkarni, Lapierre, Wallach, Murphey, Khan, and Kaufman. The population dilemma is reviewed in Newman & Matzke, Gwatkin & Brandel, Stanford, Stevens [April, 1983], Clarke, Kleinman, and Kosinski & Elahi. Medical geography is covered by Akhtar & Learmonth and Dutt [1981]. Environmental topics are treated in Karan & Iijima, Murton, and Vesilind. Recent conflicts and upheavals are covered in "Death ...," "Golden Temple ...," Gupte, "India's [Bhopal] ...," and Weisman.

The geomorphology Systematic Essay may also be used as the basis for a research paper. The chapters on physical landscapes in Greenland & de Blij are a good introduction to the subject, as are the similar chapters in a book we have cited previously--Strahler &

Strahler (see ref. on text p. 54). Butzer is a fine source for a more detailed and technical introduction, and Pitty nicely covers the subject at this level as well. Thornbury and Chorley et al. take a broader approach to this interdisciplinary earth science, and also require their readers to be familiar with the basic principles of geology.

SPECIAL EXERCISE

This exercise focuses on the dual concepts of centripetal/centrifugal forces, which should be reviewed on text pp. 491-492.

After you have clearly established in your mind exactly what constitutes each type of force, review Chapter 8 carefully and come up with an extensive list of centripetal and centrifugal forces for each of the following countries: India, Pakistan, Bangladesh, and Sri Lanka. Since the strength and cohesion of a state depends on maintaining a favorable balance of centripetal over centrifugal forces, re-examine each of these countries and "grade" them accordingly. Which are the stronger ones and the weaker ones?

Once you have accomplished your analysis, as an optional extra exercise apply this method to a number of other countries in the realms already covered in your course. Especially interesting countries in prevous chapters are Belgium, Czechoslovakia, Yugoslavia, Soviet Union, United States, Canada, Mexico, Brazil, Peru, Afghanistan, Cyprus, Iran, Sudan, just about any country in the Middle East (Israel and Lebanon in particular), Ethiopia, Somalia, Nigeria, South Africa, and Zimbabwe. In the chapters still to come, this category might include North and South Korea, China, Indonesia, Malaysia, and the Philippines. To obtain more detailed information on a country, do not hesitate to refer to the regional geographies and politico-geographical studies cited in each chapter-end bibliography (which is highlighted in the "Term Paper Pointers" section of each chapter in this Study Guide).

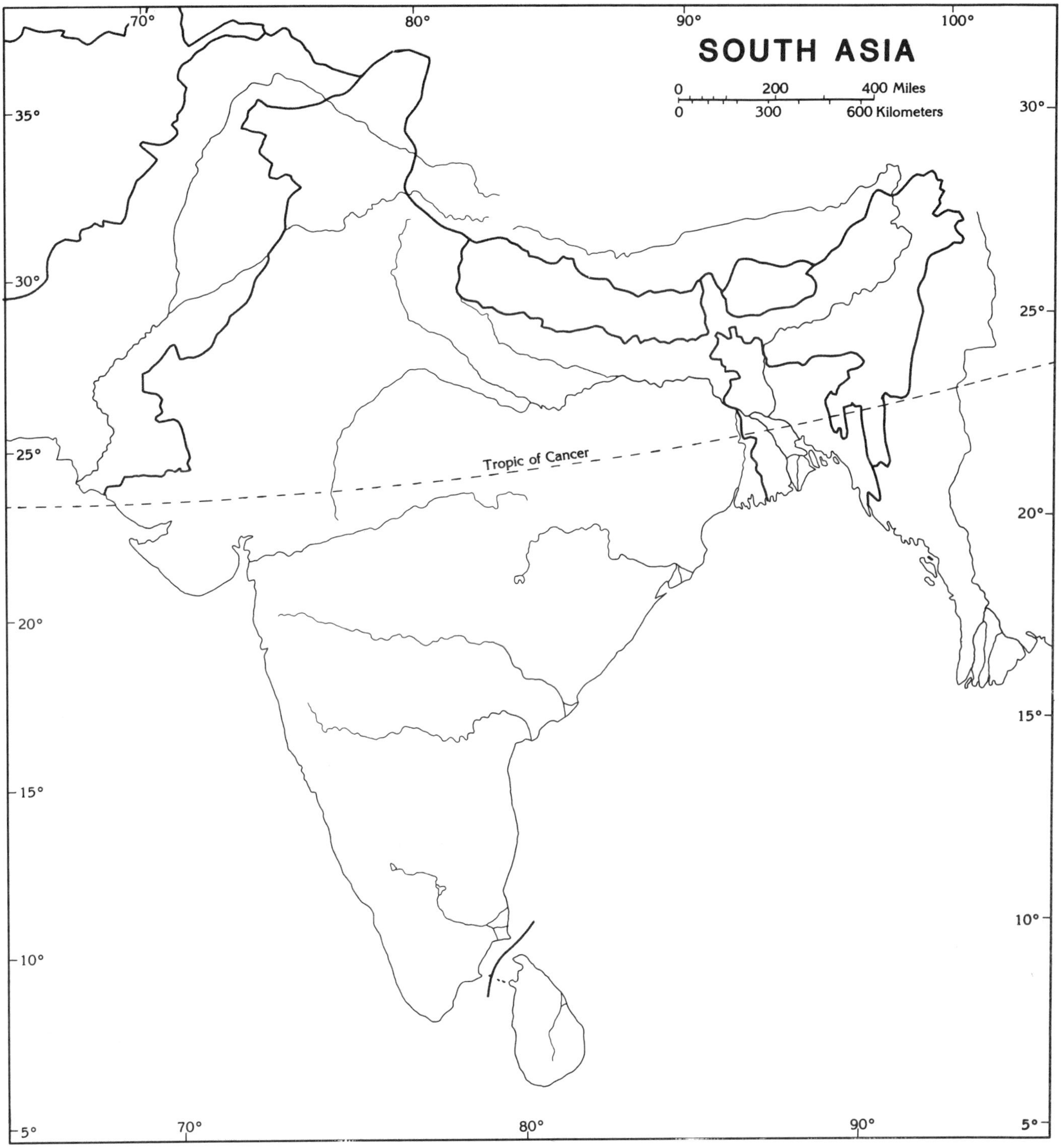

SOUTH ASIA
0 200 400 Miles
0 300 600 Kilometers
Tropic of Cancer
70°
80°
90°
100°
35°
30°
25°
20°
15°
10°
5°

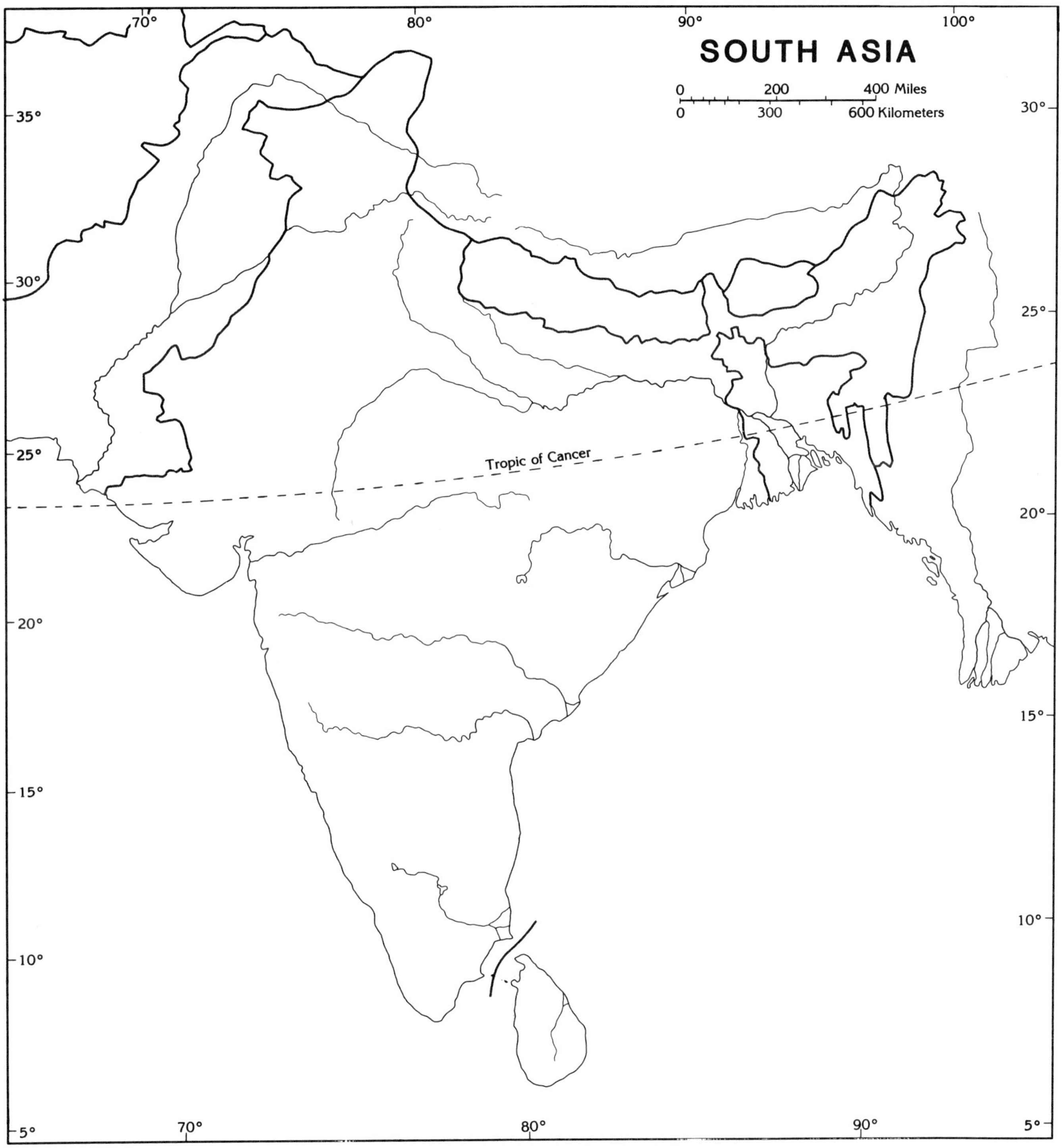
SOUTH ASIA
0 200 400 Miles
0 300 600 Kilometers
Tropic of Cancer
70°
80°
90°
100°
35°
30°
25°
20°
15°
10°
5°

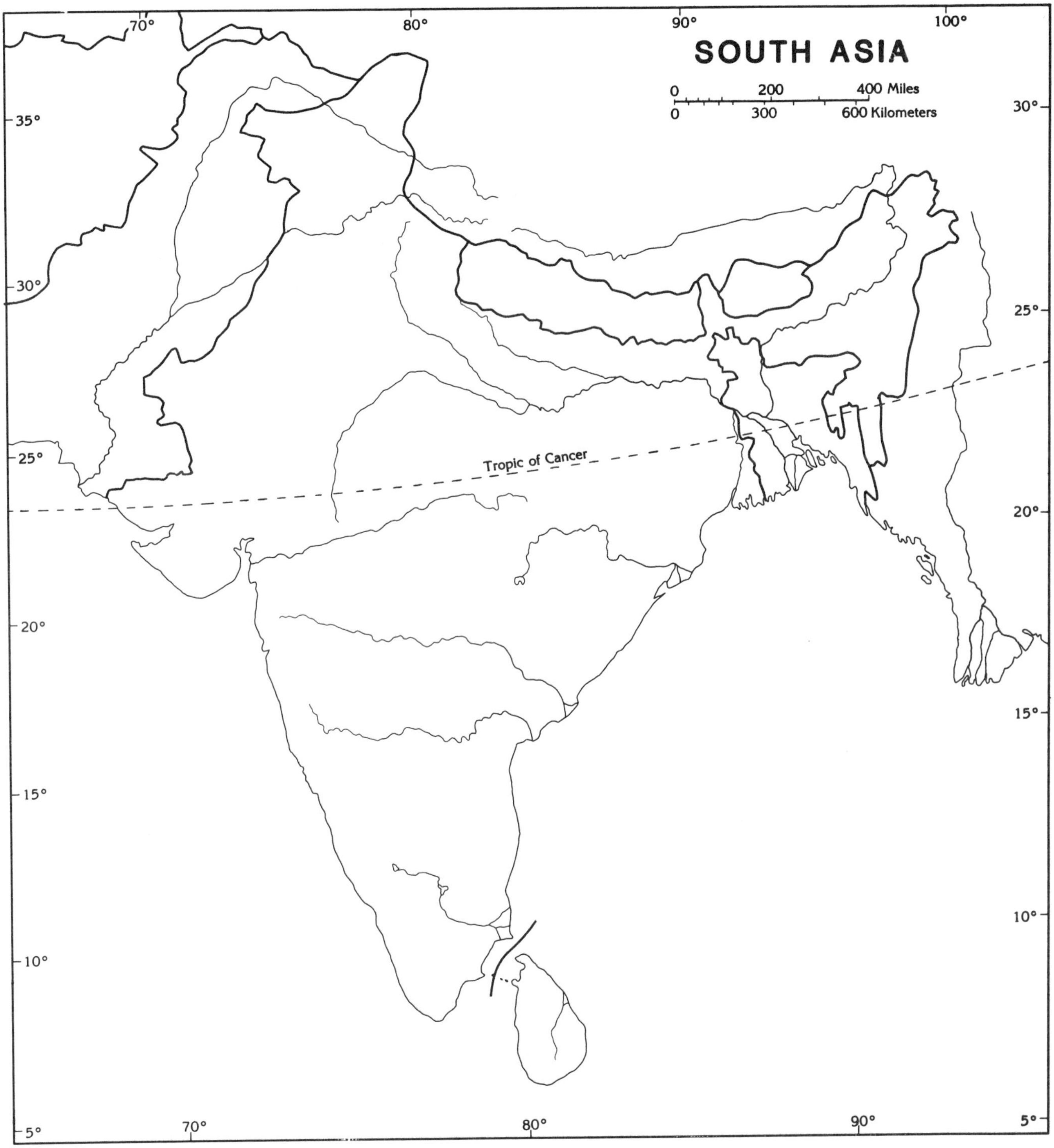
SOUTH ASIA
0 200 400 Miles
0 300 600 Kilometers
Tropic of Cancer
70°
80°
90°
100°
35°
30°
25°
20°
15°
10°
5°
30°
25°
20°
15°
10°
5°
70°
80°
90°

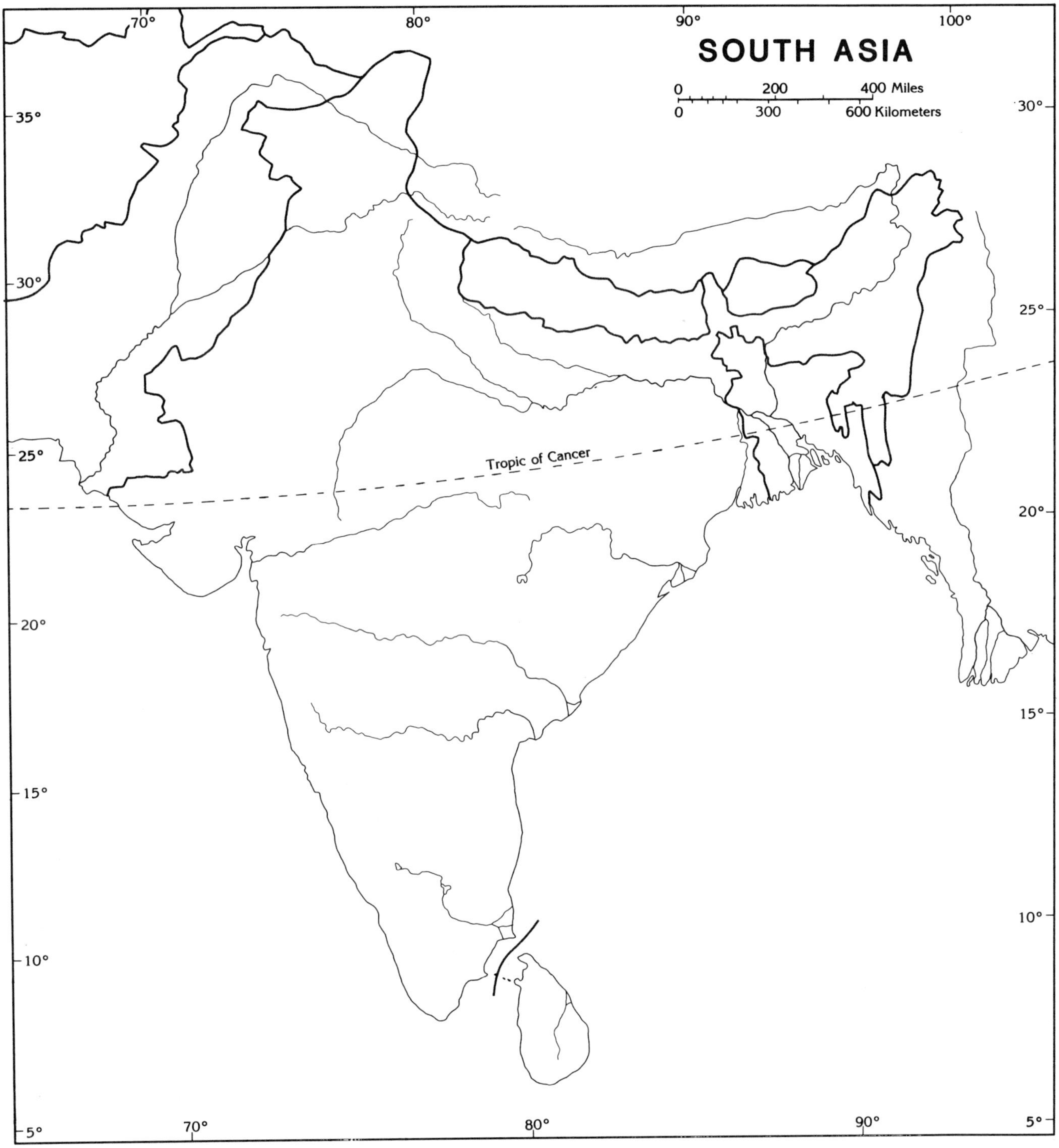

SOUTH ASIA
0 200 400 Miles
0 300 600 Kilometers
Tropic of Cancer
70°
80°
90°
100°
35°
30°
25°
20°
15°
10°
5°

CHAPTER 9: THE CHINESE WORLD -- PITFALLS OF PROGRESS

OBJECTIVES OF THIS CHAPTER

Chapter 9 covers East Asia, but mainly concentrates on China--the now awakening giant which today probably arouses more curiosity than any other world realm. Not only is China the largest country in population size and capable of becoming the third full-fledged superpower in the foreseeable future; it is also in the throes of upheaval again, but the current change promises to be a more productive and lasting one that could well propel the country into the ranks of the more modernized nations. Fittingly, the Systematic Essay here is devoted to the geography of development, once a mere branch of economic geography but now a subdiscipline in its own right that integrates several themes in contemporary human and regional geography. The introduction discusses China's current dimensions, its growing world role, and the pinyin linguistic system which is increasingly prevalent in English-language treatments of the country (and is used in the text). Historical geography follows, tracing the critical events of dynastic China, the colonial-era convulsions, and the communist transformation of the past generation. Next comes an extended regional survey, covering 4 physiographic regions and the major human subdivisions of the country (plus adjacent Hong Kong) with emphasis on the internal structure of China Proper. This is followed by a retrospect on developmental experiences since the 1949 Communist revolution, and a hard look at the ongoing Four Modernizations program now being pursued by the post-Mao leadership. Overviews of Taiwan and the two Koreas conclude the chapter.

Having learned the regional geography of China, Taiwan, and the Koreas, you should be able to:

1. Understand the basics of the geography of development and modernization.

2. Understand the pinyin spelling system as it relates to famous personalities as well as major place-names.

3. Appreciate the growing prominence and changing role of China in global affairs.

4. Appreciate the long tradition of Chinese culture and the tremendous upheavals that its adherents have experienced during the past century and a half.

5. Describe the contents and basic organization of each of China's major physical and human regions.

6. Understand the background and goals of the Four Modernizations program so that you can monitor and evaluate its progress in the future.

7. Describe the major regional patterns of two other vital parts of East Asia, Taiwan and the Korean Peninsula.

8. Locate the leading physical, cultural, and economic-spatial features of the realm on an outline map.

GLOSSARY

Center-periphery development process (518)

The most common generalized view of the spatial development process, involving changes in regional organization that result from modernization spreading outward from the major foreign contact points. Further elaborated in Opening Essay (text pp. 39-40).

Rostow economic growth model (518-520)

A framework based on the progressive absorption of technology into a national economy that occurs in five subsequent stages: traditional society, preconditions for takeoff, takeoff, drive to maturity, and high mass consumption; possibly, postindustrialism represents the most advanced (and newest) stage. Mapped by country in Fig. 9-2.

Pinyin system (524-525 box)

The standard form of Chinese established by the government in 1958 which is based on the pronunciation of the language in the north of China, increasingly adopted by foreigners (including the textbook authors) in their spellings of Chinese words and names.

Four Modernizations program (523, 554)

The ongoing development program (originally devised by Zhou Enlai in 1964) implemented by the current leadership in the late 1970s, centered around a four-pronged plan for rapid growth in industry, defense, science, and agriculture.

"People of Han" (524)

Because the Han period (202 B.C. - A.D. 220) was the formative era of Chinese culture, most residents of China still refer to themselves in this way.

Manchuria (527 box)

The foreign term for Northeast China (Liaoning, Jilin, and Heilongjiang Provinces), which the Chinese do not recognize.

Extraterritoriality (529 box)

Legal concept that the property of one state lying within the boundaries of another actually forms an extension of the first. Used in colonial-era China by European diplomats and others to carve out neighborhood enclaves that were off-limits to residents of the host country.

The Long March (530)

The exodus of the communists, who were driven out of China Proper in 1934, through the difficult Chinese interior until they had gathered sufficient strength to emerge again in the late 1940s and eventually triumph over the Nationalists in 1949.

Manchukuo (532)

The Japanese name for Manchuria during their Empire days of the 1930s and early 40s (see discussion on text p. 251 and Fig. J-3 on p. 252).

Loess (534)

Deposits of fine silt or dust laid down after transport by wind over considerable distances. These highly fertile soil-like materials can be quite productive under irrigation in the Huang He's central basin, where the Loess Plateau (downwind from the Ordos Desert) contains one of the world's largest concentrations of loess.

Qanats (537)

Underground tunnels used to transport water from mountains into nearby desert areas.

China Proper (537)

The traditional core of the Chinese nation (mapped on p. 539) south of the Great Wall and focused on the three great river basins--the Huang He, Chang Jiang, and Xi Jiang.

Great Proletarian Cultural Revolution (554)

The region of terror in the name of revolutionary zeal unleashed by Mao in the middle and late 1960s, in which the excesses of the militant Red Guards intimidated the population in the name of purging China of noncommunist traditions (and concealing the disastrous economic setbacks of the "Great Leap Forward").

Confucianism [Kongfuzi] (552-553 box)

Ideals deeply ingrained in the Chinese culture and national character, which still linger after a generation of communism.

Special Economic Zone [SEZ] (555 box)

Manufacturing and export center, established in the 1980s, to attract foreign investments and technology transfers through special tax benefits and additional economic incentives. Results have been slow in coming, and only 4 SEZs have been opened; most successful by far is Shenzhen, just across the land border from Hong Kong.

The Four Tigers of the Orient (560)

Taiwan, South Korea, Hong Kong, and Singapore--four free Asian "beehive states" which followed the Japanese route to economic success, and are now modernized and thriving.

SELF-TESTING QUESTIONS

Cover this side of the page with a sheet of paper. Uncover each line after you have attempted to answer the question in the left column. If necessary, refer to textbook page(s) listed at the right.

Question	Answer	Page
GEOGRAPHY OF DEVELOPMENT	GEOGRAPHY OF DEVELOPMENT	
What percentage of the world's population resides in UDCs?	The underdeveloped countries of the Third World realms contain over 70% of humanity.	518
What is the center-periphery spatial development process?	Organizational changes introduced by modernization spread outward from urban, foreign-contact centers toward the remote rural areas.	518
What are the 5 stages in the Rostow growth model?	Traditional society, preconditions for takeoff, takeoff, drive to maturity, high mass consumption.	518-19
What are the goals of the national development process enunciated by de Souza and Porter?	Healthful diets, good medical care, sanitation and disease control, labor opportunities, education, personal freedom, decent housing, productive activities in harmony with the habitat, a milieu promoting equality.	520
CHINA'S CHARACTERISTICS	CHINA'S CHARACTERISTICS	
How large is China's population?	1.076 billion in 1988, about one-fifth of humanity.	517
When did communism become the controlling force of the Chinese state?	In 1949, when Mao Zedong proclaimed the success of the revolution and the birth of the People's Republic.	517
Why is 1976 a year the Chinese will long remember?	Both Zhou Enlai and Mao died that year, which also witnessed the devastating Tangshan earthquake.	522
What is the pinyin system?	The standard form of Chinese, which represents a more modernized version of the language; used in the text for all Chinese names.	524

HISTORICAL GEOGRAPHY

HISTORICAL GEOGRAPHY	HISTORICAL GEOGRAPHY	
When and where did the earliest Chinese culture develop?	1800-1900 B.C. in the core area centered on the confluence of the Huang He (Yellow) and Wei Rivers.	523
What were the major accomplishments of the Han period (200 B.C.-A.D. 220), that made it the "Roman Empire" of East Asia?	Unity and stability established over a vast area (light orange shading on Fig. 9-4); trade routes secured; nomad harassments ended; private property rights emerged; strong military power; external commerce; advances in science and the arts.	524
Why was China's self-assured isolationism shattered in the 19th century?	The European powers economically destroyed China's handicraft industries, and their political influence in parts of China bordered on colonialism, a superiority enhanced by encouraging the use of opium which weakened the fabric of Chinese society.	526-29
List China's political fortunes since 1900.	1900 Boxer Rebellion began ending of colonialist meddling; 1911 overthrow of monarchy led to republican period of government led by Sun-Yat-Sen and Chiang Kai-Shek; the Japanese dominated from the late 1930s to 1946; civil war from 1945 to 1949 led to Communist takeover and Nationalist flight to Taiwan.	529-32
What is the concept of <u>extraterritoriality</u>?	The idea that a foreign power could have territorial enclaves in another country which were legal extensions of that first state.	529
What was the Long March?	The long, interior exodus of the communists in 1934-1935, who managed to avoid their Nationalist antagonists and eventually emerge victorious in the late 1940s.	530

PHYSIOGRAPHIC REGIONS

PHYSIOGRAPHIC REGIONS	PHYSIOGRAPHIC REGIONS	
What are China's three most important rivers?	Huang He (Yellow), Chang Jiang (Yangtze), and Xi (West).	533
What is the significance of the Loess Plateau in the Huang He's middle basin?	These windborne, siltlike deposits are highly fertile, a condition that persists with soil depth; intensive cultivation and dense population is the result, in an area otherwise unfavored for farming.	534

Question	Answer	Page
Where is the North China Plain, and what is its significance?	The lower basin of the Huang He, stretching as far north as Beijing; it is one of the world's most fertile, productive, and overcrowded farming regions.	534
What are the main farming/population clusters of the Chang Jiang valley?	The Red Basin or Chengdu Plain of Sichuan Province, the central basin between Yichang and Wuhan, and the lower valley concentrated around Shanghai and Nanjing.	534
What are the two main population clusters around the Xi Delta?	Guangzhou (Canton) and the still-British colony of Hong Kong.	535
HUMAN REGIONS	HUMAN REGIONS	
Describe the cultural geography of Inner Asian China.	A sparse population inhabits an arid environment; the majority are Moslems with cultural affinities across the border to residents of the U.S.S.R.'s Central Asian republics, but Sinicization is now well under way.	537-39
What is China Proper?	The China of the three great river valleys, south of the Great Wall.	539
Name China's three leading port cities.	Shanghai, Tianjin, and Guangzhou.	540
What is meant by the term <u>activity space</u>?	The daily territorial range of an individual, very limited for the Chinese peasant.	542
Which part of China exhibits settlement patterns reminiscent of <u>central place theory</u>?	The North China Plain, where the rural population is evenly distributed in a myriad of villages that lie in the regularly-spaced hinterlands of larger towns.	541-42
What is the <u>situation</u> of the capital city of Beijing?	Its excellent relative location takes advantage of proximity to the North China Plain, the Great Wall, and the port of Tianjin.	543
How does southern China differ from the North?	Water buffaloes used instead of oxen; rice and tea in the south, wheat and cotton up north; subtropical and humid in the south, dry and continental in the north; south is culturally oriented toward Southeast Asia.	544-45
What is Hong Kong's role?	Gateway to China, thriving "beehive" state, and major entrepot of western Pacific rim.	547-48

What are the leading industrial resources of the Northeast?	Iron ore and coal in the Liao Basin, fueling heavy industrial production in both Anshan and Shenyang; Daqing oilfield near Harbin.	549
THE NEW CHINA	THE NEW CHINA	
How has Chinese agriculture been organized since 1949?	Collectivization, then concentration into communes, followed by a return to a modified collective farming with private peasant plots permitted on the side.	551-53
What aspects of Confucianism [Kongzi] are still deeply ingrained in the Chinese culture?	Education, family life, political cooperation, and the course of law, literature, religion, morality, and human equality.	552-53
What are the goals of the Four Modernizations program?	Rapid modernization and mechanization of agriculture; upgrading of defense forces; modernization and expansion of industry; development of science, medicine, and technology.	554
TAIWAN	TAIWAN	
What is the economic status of this country?	Excellent: an always productive farming sector has been supplemented by a modern manufacturing sector, and this "beehive" state now uses the latest technologies.	559-60
How does Taiwan differ from the mainland P.R.C.?	Capitalist economic system, high rate of urbanization, and Western military ally.	559-60
THE TWO KOREAS	THE TWO KOREAS	
How long has the North been divided from the South?	Since the end of World War II despite the stalemated Korean War of the early 1950s.	560
How do North and South differ in their environmental conditions?	The North is drier, more mountainous, can only grow one crop annually, and contains most of the industrial raw materials; the South is moister, can support multiple-cropping and larger farming populations.	560-61
What are the human-geographic bases for reunification of this peninsula?	Common cultural background, industrial-agricultural complementarity, resource and food-supply matchings, and a trade potential on the international marketplace far beyond the mere combination of the presently-split economies.	561-62

MAP EXERCISES

MAP COMPARISON

1. The historical evolution of the Chinese state followed a spatial-expansion pattern associated with territorial gains. Trace these additions to China's core area in association with the dynastic traditions and trends that prevailed during the time of territorial acquisition. What spatial differences and similarities can you detect between the Chinese growth pattern (text map p. 526) and the expansion of Islam (p. 370), South American cultures (p. 315), North American settlement (p. 208), and the Russian Empire (p. 157)?

2. Using the map on p. 528, discuss the Chinese colonial-era experience and compare it to the geographic situation in Africa that can be traced by interpreting the maps included in Fig. 7-17 (p. 439).

3. The inset map on p. 531 briefly compares China's relative location to that of the United States. Carry that comparison further to include the natural environments of both countries, noting particularly the key similarities that exist (use the world maps in the Introductory chapter--pp. 15-25). Finally, do the same for the two population distributions, utilizing the maps on pp. 26-27 and 205.

4. Discuss the various centripetal and centrifugal forces that currently affect the Chinese state (concepts reviewed on text pp. 491-492), basing your observations on various maps in Chapter 9--especially the map of ethnic minorities on p. 545.

5. Utilizing the economic maps on pp. 542 and 544, evaluate the Chinese resource potential for attaining world superpower status in the foreseeable future. As an optional extra exercise, compare this assessment to similar ones for the United States and the Soviet Union (a suggested approach is provided in the final exercise in the Map Comparison section of Chapter 2 in this Study Guide).

MAP CONSTRUCTION

(Use outline maps at the end of this chapter)

1. In order to familiarize yourself with Chinese physical geography, place the following on the first outline map:

   a. Rivers: Chang Jiang, Huang He, Wei, Xi, Brahmaputra, Min, Mekong, Ulungur, Tarim, Liao, Songhua, Ussuri, Amur

   b. Water bodies: Lake Khanka, Sea of Japan, Korea Strait, Yellow Sea, Liaodong Gulf, Korea Bay, Bo Hai Gulf, East China Sea, Formosa Strait, South China Sea, Gulf of Tonkin, Bay of Bengal, Philippine Sea, Tai Hu Lake, Poyang Hu Lake, Dongting Hu Lake

c. Land bodies: Taiwan, Hainan Island, Korean Peninsula, Shandong Peninsula, Liaodong Peninsula, Kyushu, Ryukyu Islands, Tarim Basin, Gobi Desert, Ordos Desert, Qinghai-Xizang Plateau, Taklimakan Desert, Junggar Basin, Qaidam Basin, Red Basin (Chengdu Plain), Liao-Songhua Lowland, Yunnan-Guizhou Plateau, North China Plain

d. Mountains: Pamir Knot, Hindu Kush, Karakoram, Himalayas, Trans-Himalayas, Tianshan (Tien), Kunlun, Altun, Altay Mountains

2. On the second map, political-cultural information should be entered as follows:

a. Draw in each province-level boundary and label each province and autonomous region (use Fig. 9-8 as base)

b. Plot the concentration of China's ethnic minorities (Fig. 9-18) and learn their provincial affiliations

3. On the third outline map, urban-economic information should be entered as follows:

a. Cities (locate and label with the symbol •): Beijing, Shanghai, Hong Kong, Tianjin, Tangshan, Xian, Wuhan, Guangzhou, Luda [Dalian], Shenyang, Fushun, Anshan, Harbin, Vladivostok, Lanzhou, Baotou, Taiyuan, Qingdao, Zhengzhou, Nanjing, Hangzhou, Chengdu, Chongqing, Fuzhou, Macao, Haikou, Nanning, Lhasa, Xuzhou, Jinan, Urumqi, Yumen, Golmud, Karamay, Yichang, Guiyang, Taipei, Taichung, Hsinchu, Seoul, Pyongyang, Pusan, Taegu, Samchok

b. Economic Regions (identify with circled letter):

A - Red Basin (Chengdu Plain)
B - North China Plain
C - Liao-Songhua Lowland
D - Loess Plateau
E - Xi Jiang Delta
F - Shenzhen Special Economic Zone

PRACTICE EXAMINATION

SHORT-ANSWER QUESTIONS

Multiple Choice

1. Which movement was launched by Mao Zedong in the 1960s to rekindle enthusiasm for the Chinese brand of communism:

   a) Four Modernizations Program
   b) Great Proletarian Cultural Revolution
   c) Confucianism [Kongzi]
   d) Great Leap Forward
   e) The Long March

2. Which of the following regions is often called "Manchuria" by uninformed foreigners:

   a) Northeast China
   b) Taiwan
   c) North China Plain
   d) Xizang
   e) Red Basin of Sichuan

3. The main river serving the hinterland of Shanghai is:

   a) Xi Jiang
   b) Huang He
   c) Chang Jiang
   d) Yellow River
   e) Kongfuzi

True-False

1. China first surpassed the population total of 1 billion in the early 19th century.

2. The island-nation of Taiwan was formerly called Formosa.

3. Guangzhou is the leading Chinese city of the Xi Jiang Delta.

Fill Ins

1. The three main river valleys that define China Proper are the Xi Jiang, Chang Jiang, and _______________.

2. The leader of the legendary Long March through interior China was ___________.

3. The foreign power that established the puppet state of "Manchukuo" was _______.

Matching Question on Chinese Places

_____	1.	Tiananmen Square	A.	Guangzhou
_____	2.	Nationalist Chinese capital	B.	Nanjing
_____	3.	Dalai Lama's former domain	C.	Northeast China
_____	4.	Yichang gorges of the Chang Jiang	D.	Grand Canal
_____	5.	Pittsburgh of China	E.	Sichuan's Red Basin
_____	6.	Interior Karamay oilfields	F.	Tarim Basin
_____	7.	Sha Mian Island	G.	Xizang
_____	8.	Goal of Long March	H.	Shanghai
_____	9.	China's largest city	I.	Beijing
_____	10.	Upper basin of Huang He	J.	Hong Kong
_____	11.	North China Plain	K.	Shaanxi Province
_____	12.	Home of the Manchus	L.	Shenyang
_____	13.	Taklimakan Desert	M.	Gezhouba Dam
_____	14.	Kowloon Peninsula	N.	Junggar Basin
_____	15.	Chengdu Plain	O.	Ordos Desert

ESSAY QUESTIONS

1. Describe the Rostow economic-growth model, highlighting the developmental activities of each of its five stages. Where in this framework is China positioned today (offer as much evidence as you can to support your case)?

2. Discuss the key developments in the evolution of the Chinese state over the past 4000 years. Emphasize China's rich cultural heritage, explaining why Confucian [Kongzi] traditions still persist in many quarters despite the dominance of communism since mid-century.

3. China's far western desert basins of Xinjiang have experienced considerable growth in recent years. Why is it so strategically important for China to open up its Inner Asian territory, and what are some of the development opportunities that exist here?

4. The North China Plain is one of the world's most densely settled areas. Discuss the historical economic geography of this subregion, the structure of its settlement pattern, and the impact that communism has had on the organization of its agricultural production.

5. Coming after three decades of almost constant upheaval, the Four Modernizations program promises to provide real stability and meaningful progress. Define and discuss the goals of this program, the resources available within China for its continuity, and the chances for external help to keep the developmental momentum going.

6. It is often claimed that North and South Korea have much to gain by reunifying. Analyze the physical, cultural, and economic geographies of each state, and demonstrate the many complementarities that might successfully be translated into real modernization if reunification were to occur.

TERM PAPER POINTERS

The "Term Paper Pointers" section of Chapter 1 in this Study Guide offered suggestions about approaching research and writing on geographic realms and their components, and you may wish to consult this material if you are undertaking a report on an East Asian region.

Several good treatments of China's <u>regional geography</u> exist, and they are almost indispensable references for regional analyses. Because of the closed nature of Chinese society and the long period of inaccessibility to China by Western geographers, this literature is not always as comprehensive as similar treatments of other world realms. The most recent works are Pannell & Ma, Buchanan et al., and Tregear, with Pannell's edited volume also quite helpful. Tuan, Spencer, and Spencer & Thomas focus on China in the mid-postwar period; both Ginsburg and Cressey [both sources], although classics, are a generation out of date, basing much of their coverages on pre-communist China. Two atlases well worth examining are Geelan & Twitchett and Central Inelligence Agency. For treatment of post-1980 trends, the following are good general sources: Barnet & Clough, Baum, "Capitalism ...," <u>China, Long-Term ...</u>, "China: ...," Ginsburg & Lalor, Goldstein & Sears, Hook, Leeming, Murphey [1980; 1986], U. S. Congress, White, and Wolfgang.

The <u>systematic</u> aspects of Chinese geography are generally more up-to-date in the literature. Environmental issues are examined in Songqiao, Smil [1984], and Ma & Noble. Cultural and historical geography are treated in depth in Pannell [1983], Kolb, Fairbank, Parish & Whyte, Wiens, March, Salisbury, and Wheatley. Political geography is handled in East and Shabad; both Lattimore and Wittfogel, in major scholarly contributions, examine historical-geographical questions essential for an understanding of the political structure of modern China. Urban geography, often with a strong historical focus, is surveyed in Kirkby, Sit, Murphey [1984], Whyte & Parish, Leung & Ginsburg, Chang, Ma & Hanten, Skinner, and Murphey [1954]; another colossal work of brilliant scholarship is Wheatley, a <u>tour-de-force</u> of geography and history that is one of the most important books on Chinese urbanization ever written. Economic geography in the context of regional development is treated by Jao & Leung, Pannell [1985], Smil [1985], Wong & Chu, Wortman, and Young. Population trends of the 1980s are covered in Ma, Song et al., and Ness & Ando.

As for the remainder of East Asia, Hong Kong is surveyed by Chiu & So, Kelly, Benton, and Leung et al.; Taiwan by Ho and Knapp; and the Koreas by Bartz, Bunge [both sources], McCune, and the Time report entitled "Inside the Hermit Kingdom." The Four Tigers of the Orient (see text p. 560) are treated collectively in "Special Report ...," Benjamin & Kudrle, Conkling & McConnell, Hofheinz & Calder, Linder, Lohr, and Smith.

The Systematic Essay on the geography of development is also a fruitful departure point for a research paper. Several excellent treatments of the subject can be found, and they will quickly lead you in any direction you wish to go; you are also reminded to review appropriate references in earlier chapter-end bibliographies, especially if your focus is on another underdeveloped realm. A splendid introduction to the subject is de Souza & Porter's brief overview, supplemented by a reading of McGee. You are then ready to proceed to full-length works and case studies, represented in the text bibliography by Forbes, Simpson, Rostow, Johnson, Mabogunje, Chisholm, Cole, Charlton, Dickenson, Gore, Kresge, Muegge & Stohr, Reitsma & Kleinpenning, and Vining.

SPECIAL EXERCISE

The development theme is pervasive throughout Chapter 9, and this exercise will give you first-hand experience in working with numerical data relating to national development.

Begin by rereading the box on text p. 37 which lists several quantitative measures of comparative economic development. Your main task is to select one (or more) of these measures and compile the pertinent data for all of the world's countries that are listed in the left-hand column of the Appendix A table (text pp. 611-614). The measure you choose will depend on the availability of appropriate data in your library. Your reference librarian will be able to assist you in identifying relevant publications from the United Nations and other international statistical agencies that are on the library's shelves (always try to obtain the latest data--certainly go back no further than 1983). Once you have located your data, transformed them to the right measure (if necessary), and recorded them on your list of countries, you are ready to undertake some broad international comparisons.

Rank all of your national data from highest to lowest values and carefully study the results. Based on the Rostow model (text pp. 518-519), can you detect any "breaks" in your ranked listing that suggest the existence of any of the five economic-growth stages? What other groupings do you observe? Do they have any regional patterns to them? What observations can you make with respect to each of the world geographic realms used in this book--do they emerge as "blocs" of grouped countries exhibiting similar values, or do they contain large internal variations that disperse their figures widely in the world listing (you are reminded that the countries are grouped according to realm in the Appendix A table)? After you have recorded your observations on the statistical pattern for each realm, compare them to the world norm or average figure if such a number is provided in the data source you used (most listings will give one) and rank the realms in descending order. Wrap up your analysis by drawing conclusions, both those prompted above and any other interpretations you can make.

CHINA
0 200 400 Miles
0 300 600 Kilometers
70°
80°
90°
100°
110°
120°
130°
140°
50°
40°
30°
20°
Tropic of Cancer

CHINA
0 200 400 Miles
0 300 600 Kilometers
Tropic of Cancer
70°
80°
90°
100°
110°
120°
130°
140°
50°
40°
30°
20°

CHINA
0 200 400 Miles
0 300 600 Kilometers
70°
80°
90°
100°
110°
120°
130°
140°
50°
40°
30°
20°
Tropic of Cancer

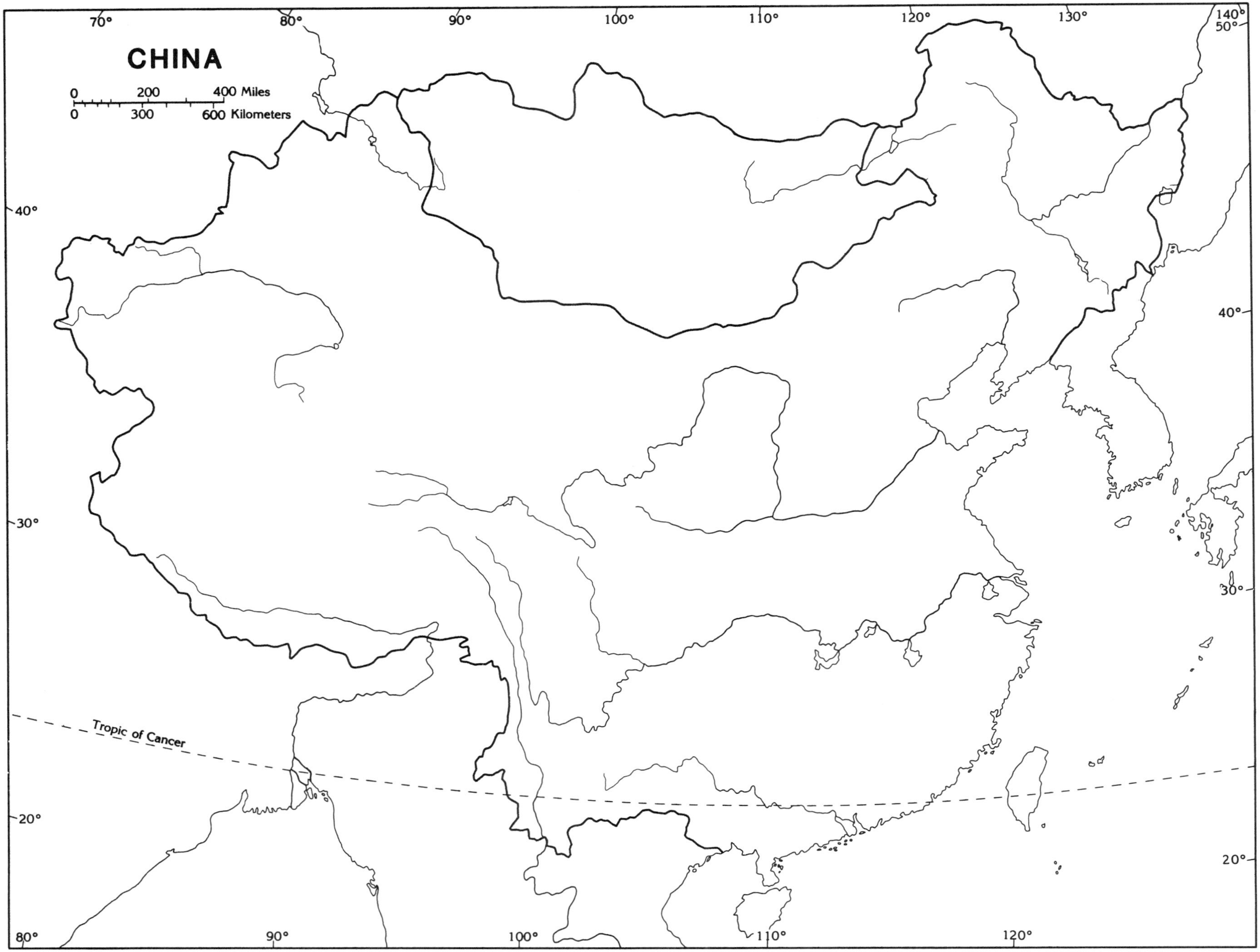

CHINA
0 200 400 Miles
0 300 600 Kilometers
70°
80°
90°
100°
110°
120°
130°
140°
50°
40°
30°
20°
Tropic of Cancer

CHAPTER 10: SOUTHEAST ASIA--BETWEEN THE GIANTS

OBJECTIVES OF THIS CHAPTER

Chapter 10 covers Southeast Asia, a realm of such cultural diversity resulting from centuries of political contests by more powerful neighbors and European outsiders that it is a classic example of a shatter belt. The Systematic Essay--political geography--is thus particularly suitable here, and the realm abounds with examples that demonstrate the territorial dimensions of political organization. Following the introduction, population patterns are treated, and major subdivisions of the realm such as Indochina are reviewed. This sets the stage for an extended look at European colonial frameworks, highlighting the experiences of the British, the Dutch, and other European countries--including a review of the Chinese experience, which in many ways parallels colonialism. Territorial morphology is discussed at length, and provides the framework for the discussion of the realm's major states. The chapter closes with a broader consideration of the increasingly important political geography of the oceans, also examining the somewhat similar approaches to the territorialization of Antarctica.

Having learned the regional geography of Southeast Asia, you should be able to:

1. Understand the basic concepts of political geography, especially the evolution of international boundaries.

2. Describe the population distributions of the realm's mainland, peninsular, and island components.

3. Grasp the various impacts that European colonialism shaped in this part of the world.

4. Describe the structure of the Southeast Asian city.

5. Understand the various categories of shape with respect to a state's territory, and the consequences of such spatial morphologies on the political development of Southeast Asia's major states.

6. Understand the political geography of maritime claims, and the partitioning of the Antarctic continent.

7. Locate the major physical, cultural, and economic-spatial features of the realm on an outline map.

GLOSSARY

Shatter belt (567)

A culturally- and politically-fragmented region shaped by the collisions of stronger and aggressive external powers, whose clashes have "shattered" the region's earlier cultural uniformity; Southeast Asia and Eastern Europe are classic examples.

Buffer zone (567)

A set of countries separating political or ideological adversaries. Thailand served as a buffer state between this realm's British and French colonial domains.

Political geography (570)

The study of the spatial dimensions and expressions of political behavior.

Territorial morphology (570)

The size and shape of a state, and what that means in national political life.

Geopolitics (570)

The application of geographic information and spatial perspectives to the development of a state's foreign policies.

Definition (570)

In boundarymaking, a treatylike document describing in words the location of an international border.

Delimitation (570)

The drawing of a defined boundary on the map with the approval of the states being divided by that line.

Demarcation (570)

The actual placing of a boundary on the landscape in the form of a fence, cleared strip, or some other physical obstacle.

Antecedent boundary (570)

One defined and delimited before the main elements and settlement patterns of the present cultural landscape began to develop.

Subsequent boundary (571)

One that developed contemporaneously with the evolution of the major spatial elements of the cultural landscape.

Superimposed boundary (571)

A boundary placed by powerful outsiders on a developed human landscape, usually ignoring pre-existing cultural-spatial patterns.

Relict boundary (572 box)

One that has ceased to function but whose imprints are still visible on the cultural landscape (such as Hadrian's Wall depicted on text p. 36).

Compact state (581)

One possessing roughly circular territory in which the distance from the geometric center to any point on the boundary exhibits little variation; Kampuchea is an example.

Fragmented state (580, 587)

One whose territory consists of several non-contiguous parts, often isolated from one another by international waters or even the land areas of other states; Malaysia, Indonesia, and the Philippines are examples.

Prorupt state (582)

One possessing territory that is at least in part a narrow, elongated land extension protruding from a more compact core area; the Malay Peninsula portions of both Burma and Thailand are examples.

Elongated state (584)

An attenuated state consisting entirely of territory that is at least six times longer than its average width; Chile is the most extreme case, and in Southeast Asia both Vietnam and Laos are good examples.

Domino theory (585-586 box)

The idea that a group of neighboring states would succumb in a sequence of communist takeovers--i.e., fall over as in a row of falling dominoes.

Perforated state (587 footnote)

One which completely encloses another state and is therefore perforated by it; most are small enclaves (such as the Vatican which perforates Italy)--the largest example is Lesotho within South Africa (inset map on text p. 581).

Archipelago (589)

A set of closely-grouped islands, often elongated into a chain.

"Culture System" (590)

The Dutch colonial system of agricultural control over Indonesia, involving forced-crop and forced-labor practices, which was designed to maximize revenues for the colonialists.

Territorial sea (596)

Zone of water adjacent to a country's coast, held to be part of the national territory and treated as a segment of the sovereign state.

Exclusive Economic Zone (599)

In a zone extending seaward for 200 miles, coastal states can control fishing, mineral exploration, and other economic activity.

SELF-TESTING QUESTIONS

Cover this side of the page with a sheet of paper. Uncover each line after you have attempted to answer the question in the left column. If necessary, refer to textbook page(s) list at the right.

Question	Answer	Page
POLITICAL GEOGRAPHY	POLITICAL GEOGRAPHY	
What are centripetal and centrifugal politico-spatial forces?	Centripetal forces unify the state; centrifugal forces are divisive.	570
What is meant by the term territorial morphology?	The size and especially the shape of a country and what they mean in national political life.	570
Distinguish among the boundary-related terms of definition, delimitation, and demarcation.	Definition is the document describing the boundary in words; delimitation is its placement upon an official map; demarcation is the actual placement of the border on the landscape.	570
Distinguish among antecedent, subsequent, superimposed, and relict boundaries.	Antecedent boundaries precede human-landscape development; subsequent boundaries emerge together with the elements of the cultural landscape; superimposed boundaries are forced by outsiders with little regard for existing cultural patterns; relict boundaries no longer function but are still evident in the landscape.	570-72
REALM CHARACTERISTICS	REALM CHARACTERISTICS	
Why is Southeast Asia a shatter belt?	Its cultural-spatial fragmentation results from the collision of stronger outside powers within the realm, much like Eastern Europe.	567
Describe the realm's population distribution. Why is it so different from the rest of Asia's?	Its few notable population clusters are fairly small and lie separated by wide expanses of sparse settlement. Physical obstacles are widespread, and agricultural opportunities are much more limited.	568-69
Name the major population concentrations of this realm.	The basins of the Irrawaddy, Chao Phraya, Mekong, and Red Rivers; the fertile volcanic island of Jawa; the plantation-studded western coast of the Malay Peninsula.	572

Describe the contents of Indochina.	The former French holdings of Laos, Cambodia (Kampuchea), and Vietnam form the core but the term well applies to the bulk of mainland Southeast Asia, underlining the two major Asian influences that have affected this realm for the past 2000 years.	574-76
Describe the distribution of the Chinese in Southeast Asia.	As the map on text p. 576 shows, there is a wide range of penetrations, peaking in western Malaysia, Jawa, the Mekong Delta, central Thailand, Singapore, and the northern portions of Vietnam, Burma, and the Philippines.	574-76
COLONIAL FRAMEWORKS	COLONIAL FRAMEWORKS	
What were the major colonies of the British, French, and Dutch?	The British held Burma, Malaya, northern Thailand, and northern Borneo; the French ruled Indochina; the Dutch dominated the East Indies (now Indonesia); as Fig. 10-8 also shows, the Spanish held the Philippines and the Portuguese eastern Indonesia.	578-80
Describe the origin of the modern state of Malaysia.	Upon independence in 1963, a federation emerged that included Singapore; however, Singapore severed its ties in 1965, and today this state consists of Peninsular Malaysia and Sarawak and Sabah on Borneo.	578
Distinguish among <u>compact</u>, <u>fragmented</u>, <u>prorupt</u>, and <u>elongated</u> state territorial morphologies.	Compact states are roughly circular in shape; fragmented states are split into two or more parts, often separated by international waters; prorupt states exhibit long, narrow extensions; elongated states are at least six times longer than their average widths.	580
List the chief <u>centrifugal</u> politico-geographical forces for the following countries: Kampuchea, Malaysia, Indonesia, Burma, and Vietnam.	Despite Khmer cultural unity, Kampuchea has been brutally treated by competing communist factions since 1975; Malaysia and Indonesia both suffer the effects of politico-spatial fragmentation which include considerable internal cultural variations; Burma is beset by the problems associated with two core areas; Vietnam was comprised of the separate culture areas of Tonkin, Annam, and Cochin China, and suffered a generation of brutal warfare (from the mid-1940s to the mid-1970s), but is now reunited under communism.	580-88

Describe the impact of colonialism on the structure of cities in this realm.	As Fig. 10-14 illustrates, a largely European sector of high-income residential areas leads away from the port, separated from the native residential areas; a suburban ring is visible as are outsider-dominated commercial and industrial zones.	590-91
How did the Dutch administer their Southeast Asian colonies?	Inflexibily, imposing their culture system of forced-crop and forced-labor practices; although this rigidity was relaxed after 1900, the Indonesian independence movement constantly expanded and finally triumphed in 1945.	589-91
What is the major religion of Indonesia?	Islam by far, comprising the world's largest Moslem country; but the faith is taken more casually here, and the islands east of Jawa include Hindus, Protestants, and Roman Catholics.	589
Why is the beehive city-state of Singapore so highly developed?	The British made it a major economic center and port, and its enterprising Chinese majority continues to thrive; manufacturing is well advanced, and today this is one of the western Pacific Basin's Four Tigers.	592-93
MODERN POLITICAL TRENDS	MODERN POLITICAL TRENDS	
Which countries have been the major recipients of recent Indochinese refugees?	As Fig. 10-12 shows, Thailand, Malaysia, Indonesia, China, Taiwan, and Hong Kong; the United States, Canada, France, Australia, and Britain have also accepted sizable numbers.	586-87
Which religion is dominant in the Philippines?	Roman Catholicism, accounting for about 85 percent of the population.	592
Why do many countries now seek to widen their maritime claims?	Because other countries do it; the desire to exclusively control offshore mineral and fuel deposits; the desire to control nearby fishing grounds.	596-97
How does this "scramble for the oceans" support the "world-lake concept"?	With the hypothetical imposition of 200-mile territorial seas and ultimate maritime claims (see Fig. 10-20), the global boundary framework is fully extended over the oceans.	598-60
How are Antarctic territorial claims organized?	As Fig. 10-21 shows, into wedge-shaped sectors.	600-1

MAP EXERCISES

MAP COMPARISON

1. The importance of the Chinese in this realm has been amply demonstrated in the text. Compare Fig. 10-6 (p. 576) to the map of Southeast Asia's core areas (Fig. 10-2, p. 569), discussing the spatial correspondence between the distribution of ethnic Chinese and the realm's key political decision-making locations.

2. The cultural diversity of Southeast Asia is readily apparent in the map of the realm's ethnic mosaic (Fig. 10-7, p. 577). Compare and contrast this map to the distribution of religions within Southeast Asia (see Fig. 6-3, p. 367), making observations about the leading regional patterns of ethnic groups and religions.

3. Review the characteristics of antecedent boundaries on text pp. 570-571. Then go back through all of the previous chapters in the text in search of other antecedent boundaries on appropriate maps. Make a list: you should be able to find numerous examples of such international boundaries.

MAP CONSTRUCTION

(Use outline maps at the end of this chapter)

1. In order to familiarize yourself with Southeast Asian physical geography, place the following on the first outline map:

   a. Rivers: Irrawaddy, Salween, Mekong, Chao Phraya, Red, Mahakam, Kapuas

   b. Water bodies: Tonle Sap, Indian Ocean, Pacific Ocean, Andaman Sea, Jawa Sea, Flores Sea, Timur Sea, Banda Sea, Molucca Sea, Celebes Sea, Sulu Sea, Philippine Sea, South China Sea, Gulf of Thailand, Tonkin Gulf, Straits of Malacca, Karimata Strait, Sunda Strait, Makassar Strait, Balabac Strait

   c. Land bodies: Hainan Island, Kra Isthmus, Malay Peninsula, Jawa, Sumatera, Borneo, Sulawesi, Bali, Timur, Flores, Sumbawa, Molucca Islands, Mindanao, Luzon, Panay, Samar, Leyte, Palawan, Cebu, Visayan Islands

   d. Mountains and plateaus: Annamese Cord, Khorat Plateau, Shan Plateau, Arakan Mountains, Pegu Yama, Dawna Range, Pegunungan Barisan, Iran Mountains, Pegunungan Muller

2. On the second map, political-cultural information should be entered as follows:

   a. Label each country and its leading components (as appropriate)

   b. Locate and label each national capital (use the symbol *)

   c. Using appropriate color pencils, reproduce the ethnic map (Fig. 10-7, p. 577) but for the Chinese instead of the purple zones on this map use the red-colored areas shown on the preceding page (Fig. 10-6)

3. On the third outline map, economic-urban information should be entered as follows:

   a. Cities (locate and label with the symbol •): Rangoon, Mandalay, Moulmein, Bangkok, Thon Buri, Chiang Mai, Songkhla, Singapore, Kuala Lumpur, Pinang, Kelang, Alor Setar, Johor Baharu, Kuching, Banjarmasin, Samarinda, Kota Kinabalu, Bandar Seri Begawan, Jakarta, Surabaya, Bandung, Padang, Medan, Palembang, Surakarta, Malang, Pontianak, Ujung Pandang, Manado, Phnom Penh, Siem Reap, Viangchan [Vientiane], Louangphrabang, Ho Chi Minh City, Hanoi, Da Nang, Hue, Loc Ninh, Haiphong, Manila, Cebu, Davao

   b. Economic regions (identify with circled letter):

      A - Irrawaddy Delta
      B - Mekong Delta
      C - Tonkin Plain
      D - Golden Triangle
      E - Sarawak
      F - Sabah
      G - Kalimantan
      H - Brunei

PRACTICE EXAMINATION

SHORT-ANSWER QUESTIONS

Multiple-Choice

1. A boundary delimited before a cultural landscape develops alongs its route is known as:

   a) antecedent
   b) demarcated
   c) relict
   d) superimposed
   e) subsequent

2. Reunited Vietnam's capital (known as Saigon before 1976) is today named after the communist leader who founded modern North Vietnam, a revolutionary named:

   a) Kuala Lumpur
   b) Kim Il Sung
   c) Ho Chi Minh
   d) Dien Bien Phu
   e) Mao Zedong

3. The largest Moslem country in the world in terms of population numbers is:

   a) Egypt
   b) Indochina
   c) Pakistan
   d) Indonesia
   e) India

True-False

1. The city commanding access to the strategic Strait of Malacca is Hong Kong.

2. The Culture System is a Southeast Asian tradition introduced by the British.

3. The dominant religion of both Indonesia and Malaysia is Islam.

Fill Ins

1. The ethnic group comprising over 75 percent of Singapore's population are the ______________.

2. Both Thailand and Burma are examples of states whose territorial morphologies can be classified as ______________.

3. The most important island of the Philippines, which contains the capital of Manila, is called ______________.

Matching Question on Southeast Asian Countries

_____	1.	Largest Moslem country	A.	Brunei
_____	2.	One of Four Tigers of the Orient	B.	Philippines
_____	3.	Northeastern neighbor of Thailand	C.	Kampuchea
_____	4.	Tagalog speakers	D.	Vietnam
_____	5.	Achieved independence in 1984	E.	Burma
_____	6.	Angkor Wat ruins	F.	Singapore
_____	7.	Chao Phraya core area	G.	Indonesia
_____	8.	Irrawaddy Delta	H.	Laos
_____	9.	Contains Mekong Delta	I.	Thailand
_____	10.	Sarawak and Sabah	J.	Malaysia

ESSAY QUESTIONS

1. The population distribution of Southeast Asia differs markedly from that of the other Asian realms. Compare Southeast Asia's population pattern to India's or China's, discussing the physical, cultural, and economic geographic forces that account for the differences.

2. Review the course of colonialism in Southeast Asia, comparing and contrasting the British, French, and Dutch experiences, and highlighting the politico-geographical patterns that the colonial era has bequeathed to the present.

3. Define and discuss the contemporary Southeast Asian city's land-use areas and overall structuring. Use a sketch map to support your answer, and analyze the role of colonialism in shaping this urban pattern.

4. Define the notion of territorial morphology and discuss its application to Kampuchea, Malaysia, and Thailand as, respectively, compact, fragmented, and prorupt states.

5. The issue of maritime territorial claims has reached crisis proportions in the 1980s. Discuss the concept of the territorial sea, the increasing desire of most states to extend their legal limits into the deeper oceans, and the possible consequences if "the scramble for the oceans" should result in the realization of the "world-lake" concept.

TERM PAPER POINTERS

The "Term Paper Pointers" section of Chapter 1 in this Study Guide offered suggestions about approaching research and writing on geographic realms and their components, and you may wish to consult this material if you are undertaking a report on a Southeast Asian region.

As always, the best way to begin is by getting hold of a good regional geography, especially for a realm whose cultural and environmental diversity is as complex as Southeast Asia's. Hill [1979], Dutt, Hill & Bray, and Fryer are the most recent surveys, and Fisher, Dobby, and Spencer & Thomas are still new enough to be useful. Individual countries are covered in Aiken, Wernstedt, Ablin, Becker, Burley, Crossette, Fryer & Jackson, Kamm, Karnow, Milne & Mauzy, Ooi, Steinberg, Stewart, Wilhelm, and the *Time* article on Vietnam in 1983 ("When will the Peace Begin?").

A number of systematic geographical studies are well worth considering. The historical-cultural aspects of this realm are treated by Broek, Samuels, Purcell, Banks, Keyes, McCloud, Mitton, Pryor, Rogge, Savage, and Wolters. Economic geography and development are covered in Spencer, Pryor, White, Akaha, Benjamin & Kudrle, Hainsworth, Hansen, Hill [1977], Kumar, Momsen, Rambo, Whitmore, and Burling. Urban geography is surveyed by Leinbach & Ulack, McGee, Armstrong & McGee, Costello, Fuchs, Thrift & Forbes, Ulack & Leinbach, Wheatley, and Yeung & Lo. Current aspects of the geography of the oceans are found in Prescott [1985], Glassner, Blake, Johnston & Saunders, Kent & Valencia, Leng, Morgan & Valencia, Richardson, Sanger, and both works by Couper; Antarctica's recent experiences and changing environments are handled in Sugden, Lovering & Prescott, de Blij, Deacon, and Zegarelli. A number of pertinent works in political geography are cited below.

The Systematic Essay on political geography can also be used as springboard for many term papers. Standard textbooks are those by Bergman, Glassner & de Blij, Kasperson & Minghi, Pounds, and Taylor. A shorter treatise is offered by Soja, and Jackson & Samuels offer a large number of classic and more recent studies in their edited volume. The more theoretical-analytical work of the 1980s is covered in Johnston, Kirby, Markusen, Pacione, and Taylor & House. The subject of political boundaries is dealt with by Prescott [1978; 1985; 1987]; buffer states are handled by Chay & Ross: the territorial state is treated by Smith; fragmentation is examined by Paddison; geopolitics is surveyed by Parker and Chaliand & Rageau; revolutions are covered in Wheatcroft; Pepper & Jenkins present various spatial views of peace and war.

SPECIAL EXERCISE

This final Special Exercise focuses on the demographic data that are presented for the world's countries in Appendix A (text pp. 611-614). You will need a set of colored pencils and three blank outline maps of the world's countries (the easiest source is simply to photocopy the world map that can be found at the end of the Introductory chapter in this Study Guide). For each map you will be working with a set of national data that range from high to low values--for 1988 population totals, for example, China is the highest (1,076.0 billion) and several countries rank at the lowest value (0.2 million). Your first task will be to divide each set of numerical values into an appropriate number of classes, which normally should be about a half-dozen: the world life-expectancy map on text pp. 428-429 is an excellent model (and, in fact, could be redrawn from the slightly newer data listed in the Appendix A table). Once your class intervals are determined, begin coloring in each country. The color scheme is your own to choose, but keep in mind that the most effective way to communicate your map's patterns is to use a darker-to-lighter color scheme corresponding to the decrease in numerical values of your map classes.

The three maps to be prepared are for each of the following data categories: 1988 POPULATION (column 5), AVERAGE NATURAL INCREASE (%), MID-1980s (column 8), and 1988 POPULATION DENSITY in either square miles or square kilometers (columns 9 or 10).

Once your maps are completed you should record your major observations for each one, explaining the mapped pattern to the best of your abilities. Then set all three maps out side by side, and make as many additional comparative interpretations as you can. Do not hesitate to call upon the entirety of what you have learned in this course about the forces that shape global and regional patterns of physical, cultural, and economic geography. After all, the ways in which people have arranged themselves in geographic space represent the totality of their adjustments to their environments--and form the most fundamental of all geographical expressions.

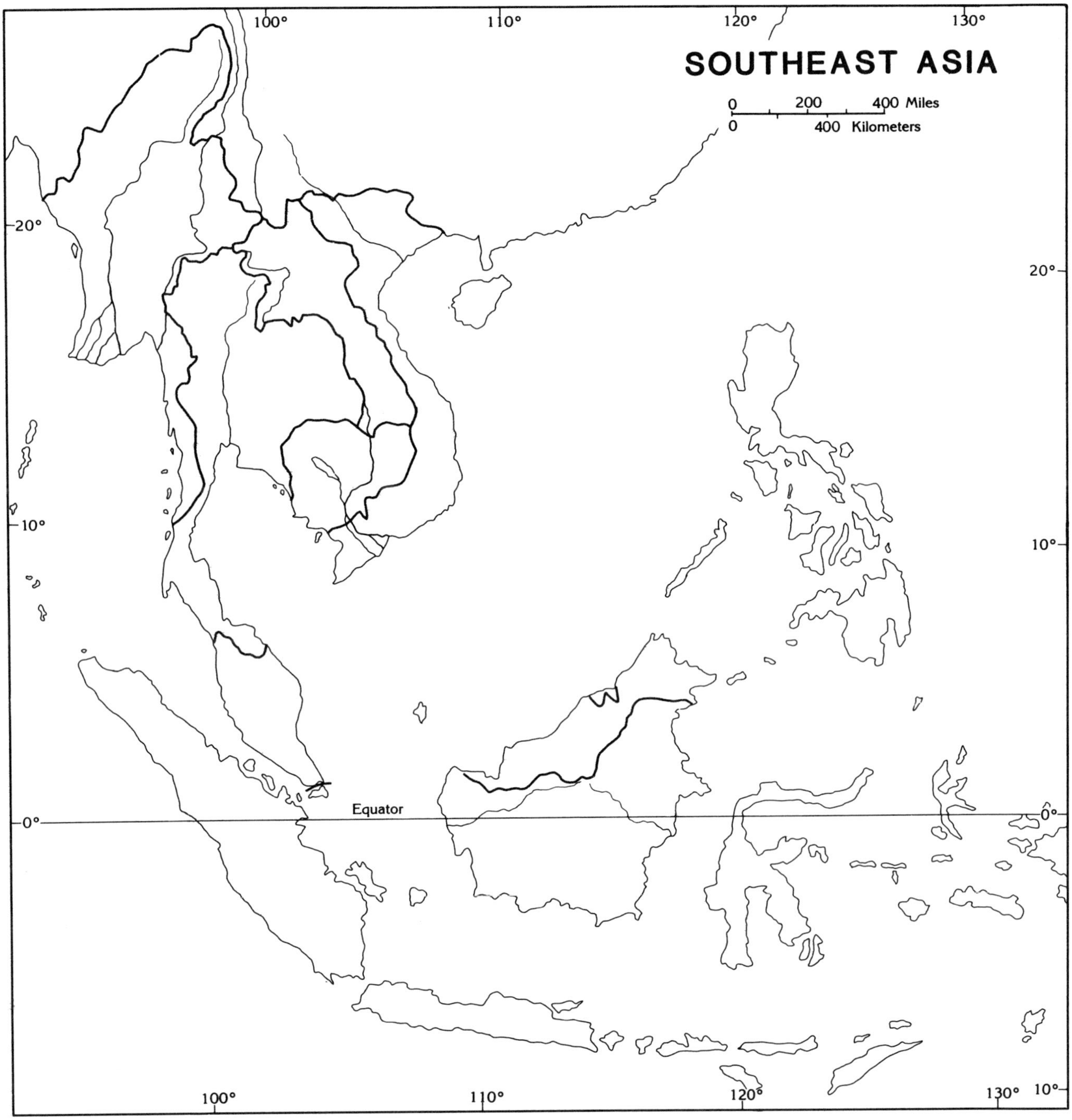

SOUTHEAST ASIA
0 200 400 Miles
0 400 Kilometers
100°
110°
120°
130°
20°
10°
0°
Equator

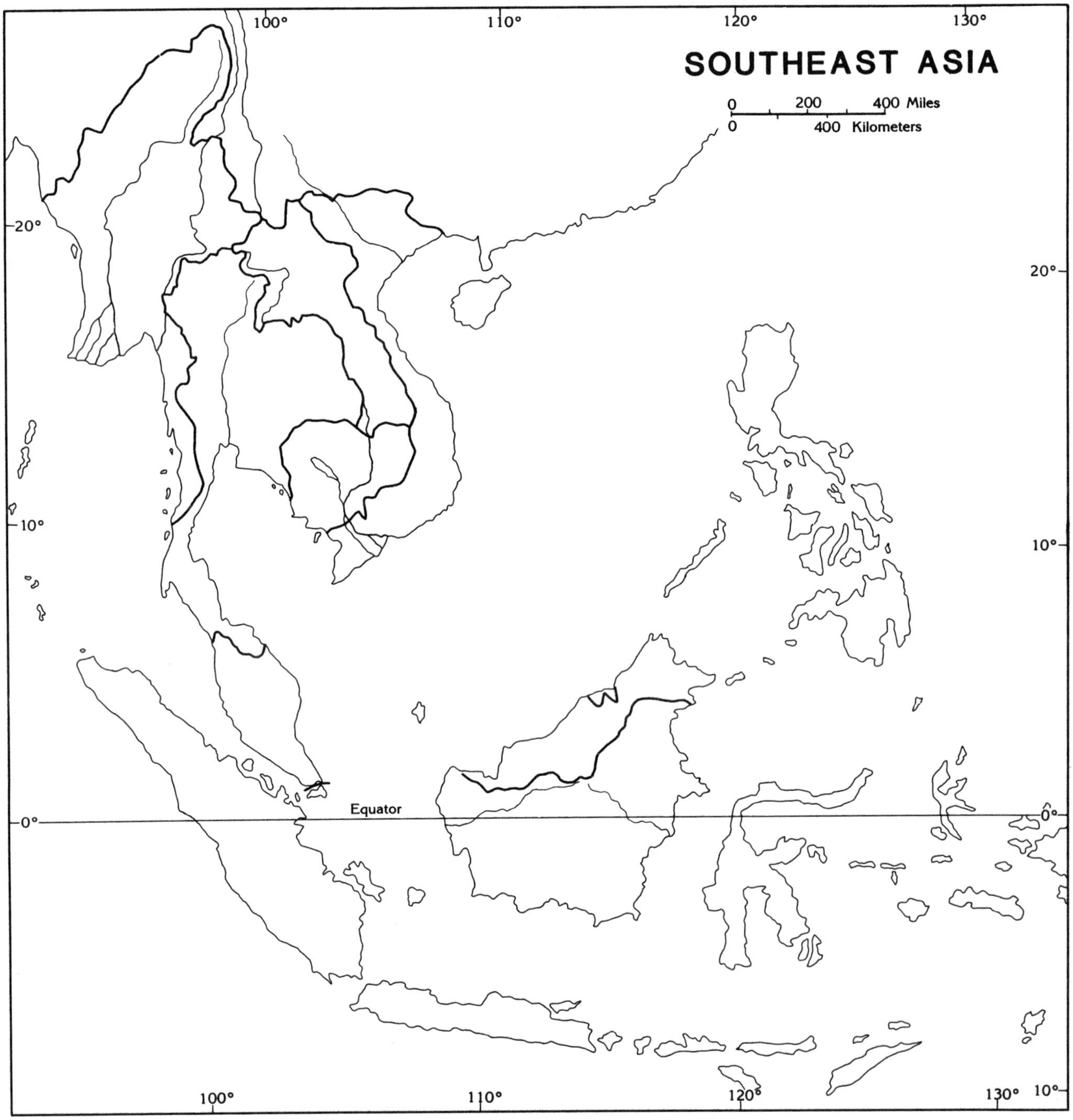
SOUTHEAST ASIA
0 200 400 Miles
0 400 Kilometers
100°
110°
120°
130°
20°
10°
0°
Equator

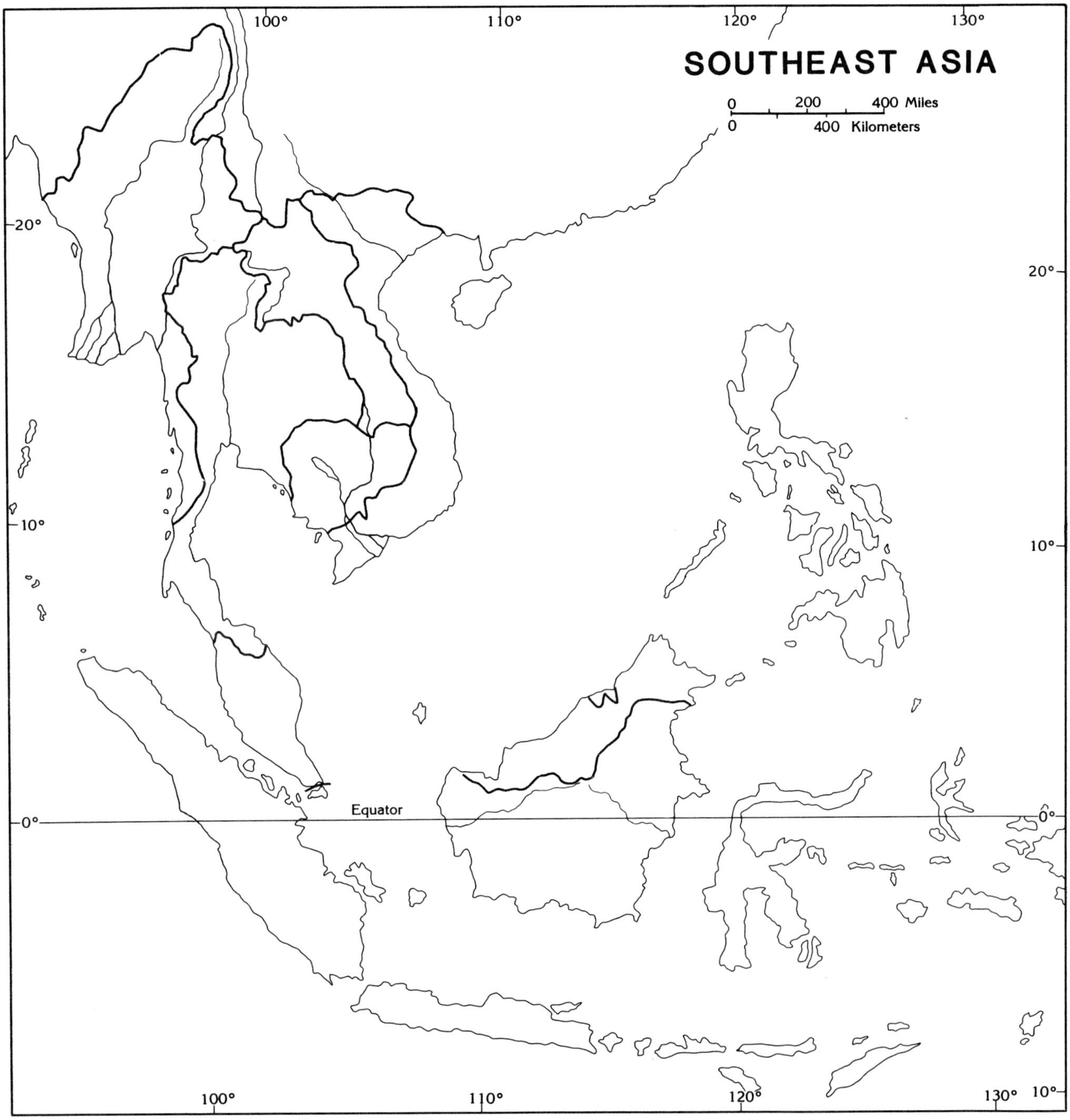

SOUTHEAST ASIA
0 200 400 Miles
0 400 Kilometers
100°
110°
120°
130°
20°
10°
0°
Equator

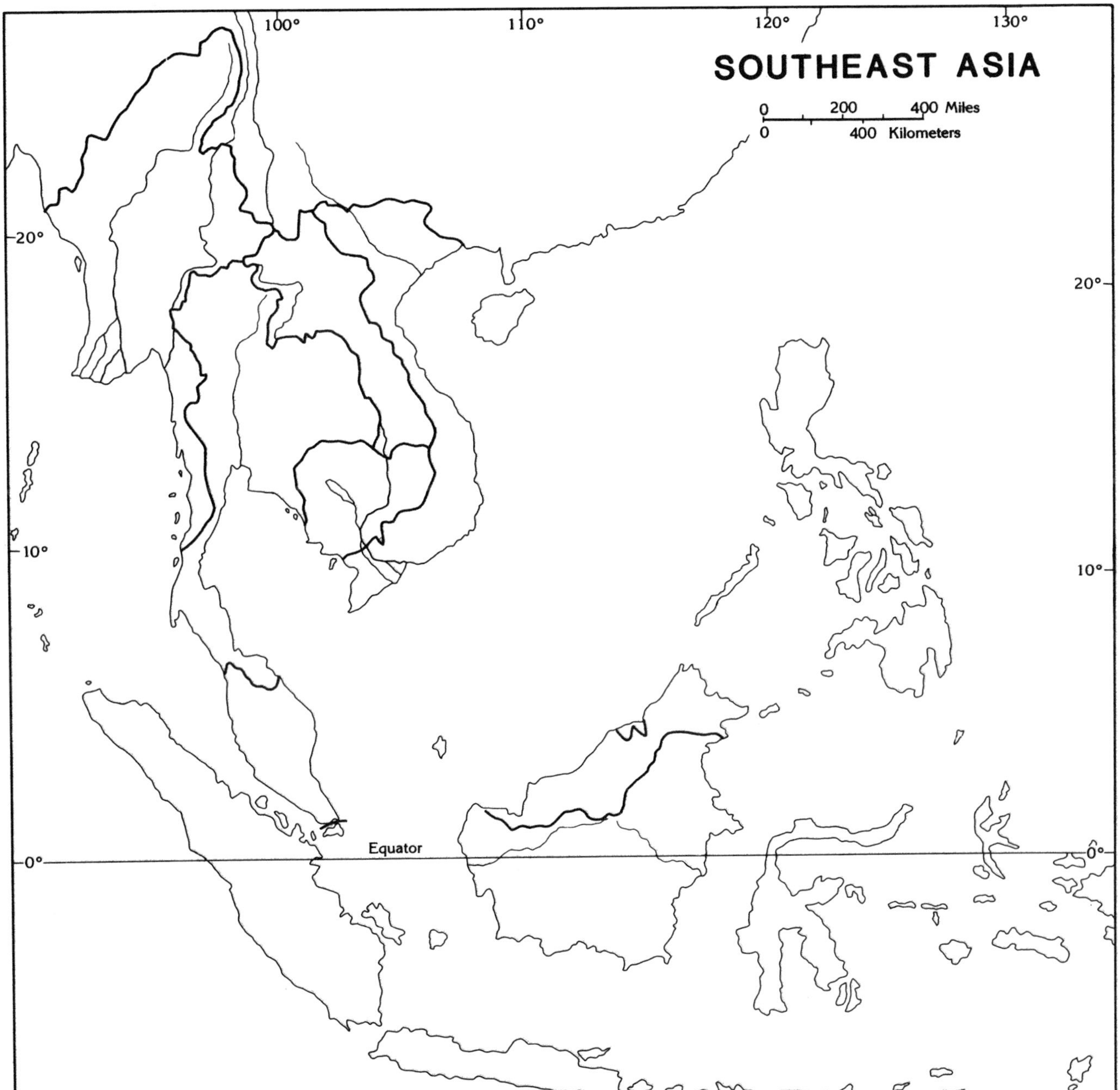

SOUTHEAST ASIA
0 200 400 Miles
0 400 Kilometers
100°
110°
120°
130°
20°
10°
0°
Equator
20°
10°
0°
100°
110°
120°
130°
10°

VIGNETTE: THE PACIFIC WORLD -- ISLAND REGIONS

OBJECTIVES OF THIS VIGNETTE

This final Vignette covers the largest area by far, greater than any of the previously-treated realms--the vast Pacific Ocean, whose thousands of islands contain inhabitants who dwell in a fragmented, highly complex geographic realm. Following a brief account of these cultural complexities, the three major regions of the Pacific Basin are identified; then, in turn, Melanesia, Micronesia, and Polynesia are surveyed. At the finish, we are deposited in the 50th U.S. state of Hawaii--thereby closing the circle as the completed survey of the earth's geographic realms returns us to the familiar developed world.

Having learned the regional geography of the Pacific realm, you should be able to:

1. Appreciate its fragmented cultural complexities as well as its regional commonalities.

2. Describe the leading geographic characteristics of Melanesia, Micronesia, and Polynesia, and locate their major places on an outline map.

GLOSSARY

Melanesia (604)

The most populous Pacific region covering New Guinea and the island groups stretching to its southeast (see Fig. P-2).

Micronesia (605)

The chains of small islands lying north of Melanesia and east of the Philippines (see Fig. P-2).

"High islands" (605)

Those of volcanic origin where some agriculture is possible; definitely in the minority among all Pacific islands.

"Low islands" (605)

Those composed of coral, low-lying so that they barely rise above sea level; usually suffer from chronic drought, and forced to rely on fishing as the subsistence livelihood; form a great majority of the Pacific's islands.

Polynesia (607-609)

The remainder of the realm, lying within the great triangle circumscribed by New Zealand, Hawaii, and Easter Island (see Fig. P-2).

SELF-TESTING QUESTIONS

Cover this side of the page with a sheet of paper. Uncover each line after you have attempted to answer the question in the left column. If necessary, refer to textbook page(s) listed at the right.

Question	Answer	Page
Why are Australia and New Zealand not included in this realm?	Their indigenous populations once qualified them as a discrete Pacific region, but they have been engulfed by Europeanization for the past two centuries.	604
What does the word Melanesia mean?	It is derived from the appearance of its indigenous peoples, who exhibit very dark skins and hair (melas means black).	604
Describe the political subdivisions of New Guinea.	The western half belongs to Indonesia (and lies outside Melanesia), forming the province of Irian Jaya (West Irian); the eastern, Melanesian half is the poverty-stricken country of Papua New Guinea.	604
What does the word Micronesia mean?	It is derived from the tiny average size of its islands (micro means small), not the physical appearance of its peoples.	605
Differentiate between "high-island" and "low-island" cultures.	High-island cultures are associated with volcanic islands whose fertile soils permit farming; low-island cultures are based on dry coral islands, very low-lying, and forced to rely on fishing.	605
What does the word Polynesia mean?	It is derived from the large number of islands it contains (poly means many) within its vast regional extent as shown on Fig. P-2.	608
How do the Polynesians differ in their appearance from the Melanesians?	They have rather lighter-colored skins, wavier hair, and are often described as possessing excellent physiques.	608
Describe the current political geography of Polynesia.	Highly complex. The U.S. 50th state comprises the Hawaiian archipelago; Tuvalu, Kiribati, and Tonga were recently granted independence by Britain; other islands remain under French control, including Tahiti; formerly German Western Samoa is now independent; other islands continue to be administered under the flags of Chile, New Zealand, and Britain.	608

MAP EXERCISE

(There are no map exercises for this Vignette)

Even though there is no outline map to accompany this vignette, the major places to learn can be listed:

Islands: New Guinea, New Caledonia, Hawaiian Is. (Oahu, Hawaii, Maui, Molokai, Lanai, Kauai), Guam, Midway Is., Wake Is., Nauru, Fiji, Tonga, Samoa, Tahiti, Easter Is., Pitcairn Is.

Water bodies: Coral Sea, Arafura Sea, Philippine Sea.

Political Units: Papua New Guinea, Solomon Islands, Vanuatu, New Caledonia, Wallis and Futuna, Western Samoa, American Samoa, Tonga, Fiji, French Polynesia, Tokelau Islands, Nauru, Kiribati, Tuvalu, Cook Islands, Easter Island, Niue, Rarotonga, Truk, and the former U. S. Trust Territory in Micronesia (the Northern Mariana Islands, Rep. of Palau, Fed. States of Micronesia, and Rep. of the Marshall Islands).

Cities: Honolulu, Hilo, Port Moresby, Suva, Pago Pago, Noumea.

PRACTICE EXAMINATION

SHORT-ANSWER QUESTIONS

Multiple Choice

1. The major economic activity associated with "low-island" cultures is:

   a) hunting
   b) farming
   c) fishing
   d) tourism
   e) nomadic herding

2. Which of the following is not a member of the South Pacific Forum:

   a) Australia
   b) Papua New Guinea
   c) United States
   d) New Zealand
   e) Fiji

True-False

1. Papua New Guinea is located in Polynesia.

2. "High-island" cultures are associated with fertile soils and farming economies.

Fill Ins

1. The most heavily populated of the three Pacific regions is ______________.

2. The largest is areal extent of the three Pacific regions is ______________.

(There is no matching question for this Vignette)

ESSAY QUESTIONS

1. Compare and contrast the geographical features of Melanesia, Micronesia, and Polynesia, and evaluate the potential for future development and modernization in each region.

2. Discuss the changing political geography of the Pacific realm, emphasizing the recent trend toward independence and the stirrings of supranationalism in the South Pacific.

TERM PAPER POINTERS

The "Term Paper Pointers" section of Chapter 1 in this Study Guide offered suggestions about approaching research and writing on geographic realms and their components, and you may wish to consult this material if you are undertaking a report on a Pacific region.

The best general survey is provided by Brookfield [1973] ; other useful overviews are offered by Alley, Spate, Vayda, Gopalakhrishnan, Kissling, Miles, "Mobility and ...," Oliver, Ward, Howells, Craig & King, and Freeman, and in the atlas edited by Wurm and Hattori. Individual regions are covered in Brookfield & Hart, Brookfield [1972], Grossman, Howlett, King & Ranck, May & Nelson, Pataki-Schweizer, Ward & Proctor, Winslow, and Howard; Hawaii is treated in Morgan, Turner, Brown, and in the revised Atlas of Hawaii. Population geography is covered by Chapman and Chapman & Prothero. A truly unique and stimulating work is the book by Levison et al., which treats Polynesian migration and settlement using the newest analytical techniques in the context of regional human-environment interaction. The latest political developments are reported in Crossette, Harden, Novak, and "Pacific Overtures."

(There is no Special Exercise in this Vignette)

ANSWERS TO PRACTICE SHORT-ANSWER EXAMINATIONS

Introduction

Multiple Choice: 1(c), 2(b), 3(a)

True-False: 1(false), 2(true), 3(false)

Fill Ins: 1(Hinduism), 2(functional), 3(B)

Matching: 1(J), 2(K), 3(C), 4(F), 5(H), 6(A), 7(I), 8(L), 9(M), 10(D), 11(E), 12(G), 13(B)

Chapter 1

Multiple Choice: 1(d), 2(a), 3(d)

True-False: 1(true), 2(true), 3(true)

Fill Ins: 1(Von Thunen), 2(Meseta), 3(density)

Matching: 1(J), 2(F), 3(N), 4(D), 5(K), 6(I), 7(C), 8(M), 9(O), 10(E), 11(L), 12(A), 13(H), 14(B), 15(G)

Australia Vignette

Multiple Choice: 1(b), 2(a)

True-False: 1(true), 2(true)

Fill Ins: 1(Wellington), 2(prison colony)

Matching: 1(B), 2(E), 3(G), 4(H), 5(F), 6(I), 7(J), 8(A), 9(C), 10(D)

Chapter 2

Multiple Choice: 1(d), 2(a), 3(b)

True-False: 1(false), 2(false), 3(true)

Fill Ins: 1(Mongolia), 2(Volga), 3(Alaska)

Matching: 1(D), 2(F), 3(A), 4(G), 5(H), 6(J), 7(B), 8(C), 9(E), 10(I)

Chapter 3

Multiple Choice: 1(d), 2(a), 3(b)

True-False: 1(true), 2(false), 3(true)

Fill Ins: 1(Boston), 2(Cascade), 3(St. Lawrence)

Matching: 1(G), 2(I), 3(F), 4(M), 5(O), 6(B), 7(D), 8(C), 9(H), 10(A), 11(J), 12(L), 13(E), 14(K), 15(N)

Japan Vignette

Multiple Choice: 1(b), 2(a)

True-False: 1(true), 2(false)

Fill Ins: 1(Kanto), 2(physiologic)

Matching: 1(D), 2(F), 3(I), 4(G), 5(J), 6(C), 7(A), 8(H), 9(E), 10(B)

Chapter 4

Multiple Choice: 1(d), 2(a), 3(e)

True-False: 1(true), 2(false), 3(false)

Fill Ins: 1(Spain), 2(caliente), 3(Mexico)

Matching: 1(O), 2(K), 3(L), 4(N), 5(C), 6(F), 7(B), 8(M), 9(E), 10(J), 11(D), 12(H), 13(G), 14(A), 15(I)

Chapter 5

Multiple Choice: 1(b), 2(a), 3(c)

True-False: 1(false), 2(false), 3(false)

Fill Ins: 1(copper), 2(CBD), 3(templada)

Matching: 1(E), 2(D), 3(K), 4(G), 5(M), 6(L), 7(J), 8(H), 9(B), 10(C), 11(A), 12(I), 13(F)

Brazil Vignette

Multiple Choice: 1(c), 2(a)

True-False: 1(false), 2(false)

Fill Ins: 1(Sao Paulo), 2(Rio de Janeiro)

Matching: 1(D), 2(A), 3(G), 4(C), 5(B), 6(E), 7(F)

Chapter 6

Multiple Choice: 1(c), 2(a), 3(d)

True-False: 1(false), 2(true), 3(true)

Fill Ins: 1(Cyprus), 2(Relocation), 3(Shi'ah or Shi'ite)

Matching: 1(K), 2(L), 3(E), 4(D), 5(O), 6(M), 7(F), 8(N), 9(C), 10(I), 11(A), 12(H), 13(G), 14(J), 15(B)

Chapter 7

Multiple Choice: 1(c), 2(a), 3(c)

True-False: 1(false), 2(true), 3(true)

Fill Ins: 1(Nigeria), 2(the Great Dyke), 3(malaria)

Matching: 1(H), 2(G), 3(I), 4(D), 5(J), 6(K), 7(A), 8(E), 9(C), 10(F), 11(L), 12(B)

South Africa Vignette

Multiple Choice: 1(e), 2(a)

True-False: 1(true), 2(false)

Fill Ins: 1(the Netherlands), 2(separate development)

Matching: 1(B), 2(E), 3(G), 4(C), 5(A), 6(D), 7(F)

Chapter 8

Multiple Choice: 1(d), 2(c), 3(a)

True-False: 1(false) 2(true), 3(false)

Fill Ins: 1(caste), 2(Coromandel), 3(physiologic)

Matching: 1(F), 2(J), 3(L), 4(H), 5(I), 6(G), 7(A), 8(K), 9(E), 10(C), 11(D), 12(B)

Chapter 9

Multiple Choice: 1(b), 2(a), 3(c)

True-False: 1(false), 2(true), 3(true)

Fill Ins: 1(Huang He), 2(Mao Zedong), 3(Japan)

Matching: 1(I), 2(B), 3(G), 4(M), 5(L), 6(N), 7(A), 8(K), 9(H), 10(O), 11(D), 12(C), 13(F), 14(J), 15(E)

Chapter 10

Multiple Choice: 1(a), 2(c), 3(d)

True-False: 1(false), 2(false), 3(true)

Fill Ins: 1(Chinese), 2(prorupt), 3(Luzon)

Matching: 1(G), 2(F), 3(H), 4(B), 5(A), 6(C), 7(I), 8(E), 9(D), 10(J)

Pacific World Vignette

Multiple Choice: 1(c), 2(c)

True-False: 1(false), 2(true)

Fill Ins: 1(Melanesia), 2(Polynesia)